PRAISE FOR
A BRIEF HISTORY OF CREATION

"Thoroughly engaging. . . . An absorbing account. . . . *A Brief History of Creation* reveals as much about the process of science as it does about the puzzle of the origin of life. That's no mean achievement."

—JOHN FARRELL,
Wall Street Journal

"A well-written and lively account of the science and history behind one of the most fascinating questions in science—how animate matter emerged from inanimate matter—enriched by engaging portraits of the scientists involved and a feel for the very human scientific enterprise at work."

—ALAN LIGHTMAN,
Professor of the Practice of the Humanities,
Massachusetts Institute of Technology,
and author of *Einstein's Dreams*

"A joyous and infinitely readable history of our ongoing quest to know how we came to be. Mesler and Cleaves elegantly narrate the evolution of philosophical and scientific inquiry, infusing the subject with all the dramatic intrigue it deserves and bringing historical figures to life as vividly as characters in a novel. A thrilling read."
—NINA SIEGAL,
author of *The Anatomy Lesson*

"With fully accessible and engaging prose, artfully weaving history, philosophy, and science, Bill Mesler and H. James Cleaves II tell what is perhaps the greatest of all scientific stories, the quest to understand the origin of life."
—MARCELO GLEISER,
Appleton Professor of Natural Philosophy,
Dartmouth College,
and author of *The Island of Knowledge*

A BRIEF
HISTORY OF
CREATION

**SCIENCE
AND THE
SEARCH
FOR THE
ORIGIN
OF LIFE**

W. W. NORTON & COMPANY

Independent Publishers Since 1923

London   New York

A BRIEF
HISTORY OF
CREATION

BILL MESLER
AND
H. JAMES CLEAVES II

For information about permission to reproduce selections from this book,
write to Permissions, W. W. Norton & Company, Inc., 500 Fifth Avenue, New York, NY 10110

For information about special discounts for bulk purchases, please contact
W. W. Norton Special Sales at specialsales@wwnorton.com or 800-233-4830

Manufacturing by QUAD Graphics, Fairfield
Book design by Helene Berinsky
Production manager: Louise Mattarelliano

Library of Congress Cataloging-in-Publication Data

Names: Mesler, Bill, author. | Cleaves, H. James, II, author.
Title: A brief history of creation : science and the search for the origin of life /
Bill Mesler and H. James Cleaves II.
Description: New York : W. W. Norton & Company, 2015. |
Includes bibliographical references and index.
Identifiers: LCCN 2015024251 | ISBN 9780393083552 (hardcover)
Subjects: LCSH: Life—Origin—Research—History. | Life—Origin—Philosophy—History.
Classification: LCC QH325.M378 2015 | DDC 576.8/3—dc23
LC record available at http://lccn.loc.gov/2015024251

ISBN 978-0-393-35319-8 pbk.

W. W. Norton & Company, Inc.
500 Fifth Avenue, New York, N.Y. 10110
www.wwnorton.com

W. W. Norton & Company Ltd.
15 Carlisle Street, London W1D 3BS

1 2 3 4 5 6 7 8 9 0

For our mothers.

Not everything mentioned in the Torah concerning the Account of the Beginning is as the vulgar imagine, for if the matter were such . . . the sages would not have expatiated on its being kept secret. The correct thing to do is to refrain, if one lacks all knowledge of the sciences, from considering those texts merely with the imagination.

—MAIMONIDES, *The Guide for the Perplexed*, c. 1190

CONTENTS

PREFACE

It was the secrets of heaven and earth that I desired to learn; and whether it was the outward substance of things or the inner spirit of nature and the mysterious soul of man that occupied me, still my inquiries were directed to the metaphysical, or in its highest sense, the physical secrets of the world.

—MARY SHELLEY, *Frankenstein*, 1818

THE SEAFLOOR WAS DARK GREEN, sloped like the outer edges of a huge jade dome, and cut by deep chasms and steep ravines. There were few signs of life. Hardly anything could be seen living that deep in the ocean, only a few slumbering giant clams and the occasional tube worm, some as large as 8 feet long. Every so often, one of the enormous worms would puff out a blood-red plume that would linger in the water like octopus ink, drifting past enormous white pinnacles that dotted the landscape. The largest towered sixteen stories above the ocean floor. Their surfaces were rough like bark, giving them the organic appearance of trees in some alien forest, spreading their branches toward the sun.

But no sunlight could penetrate that deep. A full half mile below the surface of the ocean, the ghostly pinnacles had never seen any kind of light at all until they were illuminated by the first dim flickers of a strobe light fastened to a lumbering metal craft that crept slowly above the ocean floor.

The *Argo* was about the size of a long canoe, enclosed by a large metal cage resting on sled-like rails. It didn't look like much, but the little unmanned craft already had a long and storied history. The *Argo* had taken part in some of the most important deep-sea explorations ever conducted and had found the wrecks of both the *Titanic* and the *Bismarck*. It carried some of the most sophisticated oceanographic sensors and cameras in the world, which transmitted their signals via 6 miles of cables connecting it to the surface and the research ship *Atlantis*, where geologists Barbara John and Gretchen Früh-Green steered it by means of a tiny joystick.

The scientists knew the *Atlantis* would soon have to return to port. It was early in December of 2000, and the ship had already been at sea for more than a month. The weather was turning rough—the first signs of a big squall coming in from the northeast. The *Atlantis* was a large ship—274 feet long with a crew of twenty-three and space for several times that number. It could handle a storm, but the delicate work of exploring the ocean floor required gentle seas.

The choppy water was already making it difficult to operate the *Argo*. Every swell of the ocean and sway of the ship tugged at the little craft down below. The *Argo* was tricky enough to maneuver under normal conditions, but even trickier as the scientists tried to navigate it through an undersea equivalent of the Alps, filled with unusually steep pinnacles and sharp cliff faces.

The *Argo* was exploring a deep-sea mountain called the Atlantis Massif, about halfway between Europe and North America, near the largest underwater range in the world, the Mid-Atlantic Ridge. The scientists steering from above had reason to suspect there was something special about the Atlantis Massif. The first clue was its size. It was enormous: 10 miles wide and 14,000 feet tall, about as big as Mount Rainier. The second was its composition. While much of the bottom of the ocean is covered by a mixture of minerals, the Atlantis Massif is made up almost entirely of a dense, green rock called peridotite, usually only found at least 20 miles beneath the Earth's crust. Still, John and Früh-Green didn't expect to find what no human being had ever seen, something that would become, for some, a crucial clue in the quest to answer one of the greatest enigmas in science.

As they cautiously steered the *Argo* along the rim of the mountain, the two scientists saw the first of the treelike pinnacles. More and more became visible as the craft drew closer. The first thing that struck John and Früh-Green was their height. One that would later be named Poseidon was 180 feet tall. As they steered the *Argo* toward the giant pinnacles, they noticed something else. The water was getting warmer. It dawned on them that the treelike rocks were actually what are known as submarine hydrothermal vents, a network of underwater chimneys caused by volcanic activity below the ocean floor. But these were far larger than any a human being had ever laid eyes on. When they finally spotted the first milky-white plumes of warm water coming from the tops, they realized they were looking at a new kind of chimney altogether, something few had ever guessed existed. They named the field the Lost City.

It DIDN'T TAKE LONG for news of the discovery to reach the desk of British geologist Mike Russell. Russell wasn't the only scientist in the world to suspect the existence of chimneys like those found at the Lost City. He was, however, one of the only scientists to have said with great conviction that they did exist, or at least that they had existed sometime around four billion years ago. The way Russell saw it, if those chimneys didn't exist, then we wouldn't exist, nor would any other living thing on Earth.

As a young geologist, Russell had spent several years doing fieldwork at the Silvermine Mountains in central Ireland. There he came across unique, tubelike formations of minerals that, for him, had only one possible explanation: long ago, when the whole of Ireland lay under the Atlantic Ocean, hydrothermal chimneys had percolated up through the seafloor that once existed on the dig site. What made Russell's claim so contentious was that he was talking about a kind of chimney that was different from the only ones then known to exist: the viciously hot "black smokers" that get their name from the toxic discharge of metals and sulfur that lends them the appearance of nineteenth-century industrial smokestacks. The vent that Russell imagined was merely warm, and it was rich in minerals. It was not all that hostile to living things.

Subsequent trips to the Dalmatian Alps and various mineral fields in Canada further convinced Russell that such chimneys had to exist. More important, he believed that they represented a significant clue to answering a question that had become one of the greatest challenges science had ever faced: the enigma of how life itself came to exist. Such chimneys, thought Russell, would be the perfect birthplace for life. If scientists could say conclusively where life arose, determining how it arose would be that much easier.

A century and a half before, another scientist had tried to conceive of the environment where life might have first emerged. Charles Darwin had suggested a "warm little pond." Ever since, most scientists had taken it for granted that the presence of water was necessary. They also supposed that the ancient oceans were not good candidates for the birthplace of life.

According to Russell, when the first living things appeared some four billion years ago, the oceans were about as inhospitable an environment as one could imagine. They were filled with carbon dioxide leached from the Earth's early atmosphere, leaving them too acidic to give birth to even the most primitive forms of life we know of today. They contained few of the organic chemicals that most scientists assumed were necessary for life to arise. And those chemicals would have been too dilute to easily come together to form living things.

Not so the water in Russell's chimney. The water in his "hydrothermal garden" would never get too hot or too cold. It would have been rich in a diverse variety of minerals. Also it would have been shielded from constant bombardment by meteorites and solar ultraviolet radiation on the surface world above. Despite serious opposition from veteran scientists in the field, Russell started promoting his theory with a kind of evangelical fervor. How life began was a question that had become the holy grail of the life scientists, and Russell believed he had found the key to answering it. His conviction rubbed more than a few of his peers the wrong way. To them, Russell's theory was, for the most part, mere conjecture, an interesting idea and nothing more. Supporters of hydrothermal-vent models were sometimes mockingly derided as "ventists."

Mosaic courtesy of Kelley and Elend (UW), IFE, URI-IAO, NOAA

The Lost City.

There remained one gaping hole in Russell's theory. Nothing like the environment he was suggesting had ever been found. But all that changed with the discovery of the Lost City.

IN 2009, the prestigious British science journal *Nature* ran a profile of Mike Russell. It was accompanied by a mock painting of him dressed in the dark robes and black beret of a Renaissance scientist. Perhaps the most interesting feature of the faux painting was Russell's smile. It had a certain *Mona Lisa*–like quality to it, serene and all-knowing, as if he alone held the answer to some great secret.

In less than a decade, the discovery of the Lost City had transformed a man who once sat on the peripheries of origin-of-life research into someone who could be depicted as a modern-day Copernicus in the pages of one of the world's most prestigious science journals. The interesting thing about Russell's newfound celebrity was that so many others had donned the same mantle in the past. The long saga of humanity's quest to solve the riddle of the origin of life is filled with scientists who thought they were on the brink of solving the great mystery, only to see their discoveries and contributions washed away by the acid test of scientific scrutiny. At one time or another, any one of them could easily have taken Russell's place in the *Nature* portrait.

There is something about the question of how life began that sets it apart from just about every other question science has ever tried to answer. It is not like asking how mountains form or what causes water to turn into steam. The question strikes at something deep in the very heart of human existence and at the meaning that may or may not lie behind that existence. It springs from that same intangible yearning that leads human beings to conceive of an all-powerful creator, touching upon not only how we came to be, but *why* we came to be. It is, in a sense, the ultimate question.

It takes a certain kind of iconoclast to grapple with such a question— an individual possessed of a boldness that sometimes borders on hubris. The ranks of these individuals are filled with the some of the most brilliant minds in the long history of science. Their lives have been filled with great moments of triumph and tragedy.

Their stories also teach us something about science itself. No other scientific question has ever provoked such controversy, nor has any so often stripped science of its veneer of objectivity. As much as we may wish and believe science to be the pure pursuit of truth unencumbered by human prejudice, it exists and has always existed in a world of human beings, with all of their failings and self-deceptions.

This is the story of the appearance of life on Earth. But just as important, it is the story of the evolution of how *we* see the appearance of life on Earth. From the vantage point of the twenty-first century, it is tempting to see it as a story with a clear trajectory. First there is darkness and ignorance. Gradually, this gives way to illumination and knowledge, marching onward in a straight line through Darwin's revelations of the workings of evolution, through the deciphering of the genetic code, and all the way to the unraveling of the cell's inner workings. Yet along that path, there were countless twists and turns. Ideas long discredited have found redemption; science thought incontrovertible has been disproved. More such twists will no doubt occur, for the mystery has not yet been solved. We still don't know how life began. No one was there to witness the event, and almost all of the geological record of that period has long since been erased by billions of years of constant geological change.

What we do know is that by at least three and a half billion years ago, a single-celled living organism appeared on a sterile Earth. We don't know for sure how it got there, but we can infer that it emerged from nonliving matter. An educated human being in the eighteenth century might have laughed at such a notion. Yet a person living in ancient Rome or ancient China or nearly anywhere else in the ancient world would have held beliefs not so strikingly different, at their core, from what we basically believe today. A scientist in the twenty-first century would call the emergence of life from nonlife abiogenesis. A literate Greek around the time of Christ would have called it spontaneous generation. But both understandings are, at root, quite similar. As much as it might surprise us today, throughout most of human history people didn't think that the sudden appearance of life from nonlife was all that miraculous.

A BRIEF
HISTORY OF
CREATION

BY THE ACTION OF YOUR SUN

I am above the forest region, amongst grand rocks & such a torrent as you see in Salvator Rosa's paintings vegetation all a scrub of rhodos. with Pines below me as thick & bad to get through as our Fuegian Fagi on the hill tops, & except the towering peaks of P. S. [perpetual snow] that, here shoot up on all hands there is little difference in the mt scenery—here however the blaze of Rhod. flowers and various colored jungle proclaims a differently constituted region in a naturalist's eye & twenty species here, to one there, always are asking me the vexed question, where do we come from?

—JOSEPH HOOKER, Letter to Charles Darwin, June 24, 1849

N O ONE KNEW where the river began. Its source was thought to be somewhere far to the south, beyond the distant land the ancient Egyptians called Nubia. By the time it reached Egypt, the river grew at points to more than 4 miles wide, cutting through the rocky lands that bordered the southern edge of the kingdom and forming a deep canyon some 650 miles long. Then the river hit the great Sahara desert, winding through it like a life-giving road that split the endless sands in two before settling, finally, in the Mediterranean Sea.

The Egyptians didn't have a name for the Nile. There was no need. The river was life, and all life centered on the river. They called it, simply, *iteru*, "the great river." Their country they called *Kemet*, "dark land." It was the same word they used for the abundant black soil that collected on the river's banks, deposited there by a massive annual flood. Every year around

July, *iteru* would rise, flooding the plains. Two weeks later, it would recede, leaving behind the nutrient-rich *kemet*. The size of the flood would foretell a year of abundance or famine, life or death.

And every year, as if by clockwork with the rising of the Nile, the frogs would come—thousands upon thousands of them. They were the same frogs that would inspire the story of the ten plagues of Exodus. Just as the Egyptians wondered about the source of the great river, they wondered about the frogs. As far as they could tell, the frogs didn't come from eggs as did the ibises so commonly seen nesting in the tall reeds along the river. Nor did they emerge from the womb of a mother like the water buffalo that grazed on the river's edge. In the eyes of the Egyptians, the frogs simply arose from the waters, a gift of the frog-headed fertility goddess Heket, who was said to swim the Nile during its rising.

There was nothing particularly unusual about the way the ancient Egyptians viewed the frogs of the Nile. Some creatures sprang from a mother's womb. Some hatched from eggs. Some living things just appeared, naturally, from things that were not alive. As far as human beings could

Ancient Egyptians confront the plague of frogs.

tell, those creatures simply emerged from wood or from old grain, from water or from dust. They could see this phenomenon happening all around them. Insects appeared to spring from fallen trees. Mice appeared in grain. Frogs simply emerged from freshwater.

For the ancient Egyptians, the emergence of life from nonlife was no more wondrous than a chick's emergence from an egg. That same belief in a commonplace relationship between life and nonlife shaped the way humans thought about the emergence of the first of things, whether the first chicken, the first owl, or, most important, the first human being. People found the explanation all around them, in all the eggless and parentless creatures they saw in their world.

The creation stories of most religions are remarkably similar in this regard. In the beginning, there is nothing, or at least something close to nothing. For the Hindus, it was an unknowable chaos; for the Chinese, a formless Dao. The Egyptians believed, understandably, that the universe began with only a mass of water, called Nu, which was surrounded by darkness. These formless beginnings are typically followed by a divine process of creation, culminating in the appearance of human beings, often from a natural substance that makes some cultural sense. In Egypt, the original god, Atum, spawned the rest of the gods by masturbating, or, in some versions, making love to his shadow, which earned him the title "the great He-She." Finally, from the god Ra's tears, human beings emerged—in other words, from water, just like the frogs. For the Norse, the first humans were forged from ice. The Mayans believed human beings emerged from wet clay, as did the ancient Assyrians. In the book of Genesis, "the Lord God formed man of the dust of the ground." All of these accounts must have seemed reasonable to their creators. If a frog could simply be brought to life from something like water, why couldn't a human being?

It would be wrong to see these creation myths as merely stories. They reflected the laws of nature as observed by the peoples who adopted them. There is a reason the Norse saw importance in ice, while the desert-dwelling Egyptians centered their story on water. The problem with these accounts is that they were an end unto themselves. They were unchallengeable. As knowledge of the world grew, such answers could not change.

But there was another way of approaching the question, another way of approaching all questions. This method did not seek to produce an answer, but to produce a conjecture, a hypothesis. This was not truth, but a seed of truth that would, subjected to scrutiny and critical thinking, bloom into a better understanding of the questions that humankind sought to answer. It assumed no role for a divine creator and was based solely on careful observation and deduction. Eventually, experimentation would be added to that list, but this was very early in the history of what we would now call science.

IN THE SIXTH CENTURY BC, about two hundred years after the *Iliad* was composed by the blind poet Homer, the earth shook along the Taygetos mountain range that rimmed the Greek city of Sparta. It was an earthquake so massive that, according to the Roman historian Cicero, one peak was "torn away like the stern of a ship in a storm," crushing the city below and leaving it in ruins. The Spartans themselves, however, were not harmed. In Cicero's account, they had spent the night in the fields of the valley below after being warned by a philosopher who had come from the Anatolian city of Miletus. His name was Anaximander.

The story of Anaximander saving Sparta is almost certainly just a legend. A different account places Anaximander in Sparta setting up a gnomon, a metal rod that served as an early sundial, and makes no mention of the earthquake.* To most ancient Greeks, both stories would have been equally credible. For them, the leap from keeping time to predicting an earthquake wasn't such a huge one. Both must have seemed like magic, just as reading and writing would seem like magic to peoples that didn't possess the technology.

---

* Using little more than trigonometry and sundials, Greeks were able to make accurate estimates of the Earth's circumference and shape. The philosopher Eratosthenes estimated the Earth's circumference to be about 25,000 miles. Modern measurement using satellite-based instruments has determined the circumference to be 24,860 miles.

Anaximander was born at an auspicious time in an auspicious place. By the year of his birth in 611 BC, the Greek city of Miletus had become, for a brief spell, one of the greatest city-states in one of the world's greatest empires. Miletus was ideally situated for trade, located on the coast of the southwest corner of the Anatolian peninsula in modern-day Turkey, and near the mouth of the river Maeander, with so many twists and turns that it spawned the word "meander." The Miletians were sailors of considerable renown, and the city's harbor on the Aegean Sea was always full of trading ships picking up wine and olive oil extracted from the fertile groves that dotted the countryside, or delivering shipments of a Phoenician sea snail called murex. From the murex, the Greeks extracted a valuable purple dye worth its weight in silver. As many as 12,000 snails could be used to produce a single garment. Murex-dyed cloth became associated with wealth, and the color purple became synonymous with royalty. The wealth of Miletus was legendary. The Greek historian Herodotus dubbed the city "the Jewel of Ionia."

Miletus was also a regional military power, with as many as ninety colonies of its own. Anaximander himself was sometimes described as the leader of a colony on the Black Sea. But the reason the city is still remembered has nothing to do with its money or soldiers: Miletus was the birthplace of Greek philosophy.

It was the home of the philosopher Thales, a contemporary of Anaximander and probably his mentor. Thales is remembered as the first of the Greek philosophers, which is true, and the world's first mathematician, which is not. He was also the first individual known to be credited with a mathematical discovery, that of trigonometry. In reality, trigonometry probably originated in Egypt, where, as a young man, Thales had studied the pyramids while receiving instruction in Egyptian theology.

The Greeks have always received a lot of credit for things that began with other cultures, especially those in their immediate vicinity, like the Egyptians and, even more commonly, the Babylonians. The Babylonians were, as far as we can tell, the first to really take note of the world around them, recording their observations on reddish-hued clay tablets that they baked in the sun on the banks of the Tigris and the Euphrates. They were the first to keep track of time, observing, carefully and systematically, phenomena like the speed of the sun as it bounded from horizon to horizon. They categorized all manner of heavenly events, and made huge strides in mathematics. Just as our numeric system revolves around the number 10, the Babylonians' central figure was 60, which is why we think of time in units of 60, as in 60 seconds in a minute or 60 minutes in an hour. The term "science," referring to a systematic practice of solving questions, wouldn't be coined until much later. Identifying its beginning is tricky. But one could do worse than to trace its origins to the Babylonians. A stronger case can be made for them than for the Greeks.

The Babylonians were the true inventors of the sundial, something for which Europeans would long credit Anaximander. And though Anaximander has often been called the first mapmaker, the Babylonians had already done that too, as had many other cultures. Anaximander was, at one time or another, probably falsely credited with as many inventions as any man in history. Yet he probably justly received credit for one important invention, one that would ensure he would be remembered long after all his contemporaries. Anaximander was, as far as we can tell, the first to write his thoughts down in prose in something we would call a book. He titled it *On Nature*.

On Nature was an attempt to create a total cosmology of the universe, from beginning to end. Others of Anaximander's contemporaries had imagined such cosmologies. His mentor, Thales, had a cosmology of sorts. But Thales's worldview wasn't very different from that of his countrymen. According to Thales, in the beginning there was water. The concept was something like that of Nu in Egypt, a country that Thales had visited and studied in. Thales's vision was rooted partially in observation: there is a lot of water on Earth, living things are largely composed of water, and water has the power to transform matter, such as dirt into mud. Thales believed that all things were formed of water, that it was the essence of all matter. For this, he has long been credited as the first to identify what a chemist would call an "element," a substance that cannot be broken down into components and that forms the building blocks of more complex forms.

Though we now know that water can be broken down further into the elements hydrogen and oxygen, it was a perceptive observation.*

Yet Thales's cosmology was anchored in the divine, with soul-like mystical forces that brought things to life. Anaximander's universe was different. It was rooted entirely in things he could see, and he had never seen a soul. There was no room for the mystical or the supernatural. Anaximander's universe was like a machine that could run on its own.

Anaximander sought answers about things he could physically see, the things he could truly understand. He set his sights on the sun, the stars, the Earth, and its creatures. Everything in nature, he believed, could be explained in terms of four basic elements: earth, wind, fire, and water. He supposed the sun to be twenty-eight times as big as the Earth. The real figure is about five hundred thousand times larger, but at the time it was remarkable even to recognize the sun as the larger body, given how small it appears in the sky. Anaximander thought the Earth was curved, and shaped like a stone column. The stars, he believed, traveled in great circles around the Earth. It was from this deduction that he made his most revolutionary observation of all: If the stars traveled around the Earth, then it would follow that the Earth was floating freely in space. There was nothing underneath it—a fact that would have been unimaginable to most people for the majority of human history. Even though Anaximander's conceptions were often crude, in many ways he understood the natural world at least roughly as well as any human being over the next two thousand years.

Anaximander's explanations of the development of life on Earth were just as prescient as his observations of the heavens, even if he was a little vague about how the world began. His universe began filled with a kind of infinite nothingness that he called *apeiron*. Gradually, the basic

---

* About the same time, the Indian philosopher Kanada described material building blocks, which he called *anu*, from which all matter was constructed. *Anu* were tiny (smaller than a speck of dust), indestructible spheres. The Greek philosopher Democritus reached similar conclusions and coined the term "atom." Both came to their respective theories by asking themselves the same question: If you break something in half and then continue breaking those halves into smaller halves, how long until you run out of things to break?

elements—earth, wind, fire, and water—took shape, and they began to combine to form new things. Eventually, the first plants and animals rose out of the mud where the sea meets the land. These first life-forms were initially enclosed in something like floating tree bark, which drifted until the tides carried them to the shore. Marooned, they dried in the heat of the sun, became brittle, and cracked open, setting free the creatures and plants they contained.

In Anaximander's scheme, human beings were among the first creatures to appear on the Earth, emerging out of the mouths of fishlike beasts that crawled onto the beaches, like butterflies from cocoons. In a sense, they evolved, and his narrative resembled one that might be given by a child of the twenty-first century asked to describe the process of evolution.

Anaximander must have thought the question of the origin of life rather simple. Unlike the sun, which he couldn't measure, or the stars, whose trajectory he could but guess, Anaximander could *see* living things appearing out of nonliving things. It was simply a change of form of the natural elements, like wood turning to fire, or fire to smoke. Yet he never gave the phenomenon a name. That would be done by one of Anaximander's intellectual inheritors, Aristotle.

ONLY TWO ENIGMATIC SENTENCES of Anaximander's writing survive in the form in which they were originally written: "All things originate from one another, and vanish into one another according to necessity. They give to each other justice and recompense for injustice, in conformity with the order of Time." The fragment from *On Nature* was saved by the Greek philosopher Simplicius and included as a quotation in his commentary on Aristotle's great work, *Physics*. Everything else we know about what Anaximander believed comes secondhand from references by the many learned Greeks who read his words, particularly in the work of Theophrastus, one of Aristotle's most important contemporaries.

More than two hundred years after Anaximander's death, Theophrastus likely familiarized himself with *On Nature* in Aristotle's library, the Lyceum of Athens. The Lyceum had existed long before Aristotle, but

Aristotle transformed it from a gymnasium that trained athletes for the Olympic Games into an academy for training the greatest young minds of his generation. Would-be philosophers flocked to Athens to the court of Greece's preeminent intellect, a position Aristotle inherited from his Athenian teacher, Plato. Just as Aristotle had been a student of one of history's most accomplished thinkers, he, in turn, became tutor to one of history's greatest conquerors, Alexander the Great. And as Alexander's conquests grew, so did the splendor of Aristotle's Lyceum. In addition to his famous library, Aristotle built a botanical garden and a zoo filled with animals from faraway lands, sent back by Alexander from the lands he had conquered.

Aristotle's rise to his position as Plato's intellectual heir was roundabout, and owed much to politics. The most complicated part was that he was not a Greek at all. When he was only eighteen, Aristotle had come to Plato's Academy from Macedonia, where his father was the court physician to the king, Alexander's grandfather. Aristotle's brilliance was obvious, yet when Plato died twenty years later, Aristotle entered a self-imposed exile. Macedonian armies were snatching up Greek cities one by one, and anti-Macedonian feelings ran deep in Athens. Aristotle's heritage had by then become a liability.

Aristotle left Athens for the Anatolian Greek city of Assos, just north of Miletus. Eventually, he ended up on the island of Lesbos in the Aegean Sea, where he was met by a friend from his days at Plato's Academy: Theophrastus. Theophrastus was a native of Lesbos. His given name was Tyrtamus. Because of his eloquence, Aristotle had given him the nickname by which he would later be remembered: Theophrastus, meaning "one who speaks like a god." Like Aristotle, Theophrastus had wide-ranging interests, but he devoted most of his time to studying the natural world, and botany in particular. Theophrastus wrote two important books on the subject: *Enquiry into Plants* and *On the Causes of Plants*. He would become widely recognized as antiquity's greatest authority on flora.

Aristotle's interests were more deeply rooted in the animal kingdom. Teeming with all kinds of animal life, shaped by centuries of unique evolution, Lesbos must have been, for Aristotle, a little like the Galápagos would later be for Darwin: an isolated ecosystem that provided the perfect microcosm for studying the mechanisms of nature. Just as the Galápagos would provide the observational framework for *On the Origin of Species*, Lesbos was Aristotle's inspiration for three volumes of naturalism that would eventually secure his place as the founder of biology.

Of all Aristotle's works in what would one day be called the sciences—mathematics, geology, physics—none had more of a lasting impact than his biological observations. Even though he made some serious mistakes—he believed, for instance, that women had more teeth than men—he was still such a keen observer and careful cataloger that nearly two millennia later, learned men could still read his works, not merely as quaint antiquated musings, but as exemplars of the world's finest naturalistic thought right up until the Renaissance. The Europeans of the next two thousand years looked at his observations in much the same way they looked at the colossal buildings of the Romans—in wonderment at the lost knowledge that had built the things that they, the inheritors, could not.

In his seminal book *History of Animals*, Aristotle laid out a vision of the animal world not unlike the "tree of life" that would one day be constructed by evolutionists, with long rows of species gradually evolving into humans. Had Aristotle made such a drawing, his animal kingdom would have looked much the same, with each species differing only slightly from those at either side. But Aristotle believed—as his master Plato had believed—that species, like the universe they inhabited, did not change. Nature was perfect. By all rights, Aristotle, armed with his catalog of natural observations, should have been the first evolutionist. Yet even with his ample evidence, he never guessed at the answer that many of his fellow philosophers—most famously Lucretius and Epicurus—stumbled upon.

Still, Aristotle shared the belief held by Anaximander and most others that life could naturally arise from nonlife. Just as Aristotle found plants such as mosses that could propagate without seeds, he also found animals and insects that did the same. He called the process "spontaneous generation," a phrase he first used in *History of Animals*:

Now there is one property that animals are found to have in common with plants. For some plants are generated from the seed of plants, whilst other plants are self-generated through the formation of some elemental principle similar to a seed.... So with animals, some spring from parent animals according to their kind, whilst others grow spontaneously and not from kindred stock; and of these instances of spontaneous generation some come from putrefying earth or vegetable matter, as is the case with a number of insects, while others are spontaneously generated in the inside of animals out of the secretions of their several organs.

Of all the writings by Greek philosophers about spontaneous generation, Aristotle's concept was the one that would be most remembered. As the Greek world gave way to the Roman, and the Roman to the Christian, Aristotle's theory of spontaneous generation was kept alive in the works of one of the most influential Christian thinkers of all time.

IN THE YEAR AD 415, a mob of Egyptian Christians dragged a woman named Hypatia from her home, stripped her naked, and paraded her through the streets of Alexandria before finally beating her to death with tiles. Hypatia was a mathematician who taught classical theories to anyone who would receive them, whether Christian or pagan. She identified herself as a Neoplatonist, adherent to a revivalist movement centered around classical Greek philosophy. Her public stoning was the result of an uprising against what the mob felt were irreligious actions taken by the city's prefect, Orestes. The seventh-century bishop John of Nikiû later explained that Hypatia "was devoted at all times to magic, astrolabes and instruments of music and she beguiled many people with her Satanic wiles." For modern chroniclers of the early church, the death of Hypatia would come to be seen as a turning point when people in the Western world began to question the wisdom and learning of the Greeks.

The Christian Bible contains many warnings against being swayed by secular learning. In the apostle Paul's First Epistle to the Corinthians, he states, "Be on your guard; do not let your minds be captured by hollow and delusive speculations, based on traditions of man-made teaching centered

on the elements of the natural world." Hypatia's fellow North African Tertullian, sometimes known as the father of Christian theology, echoed this
antiscientific theme in his own writings. He recalled a famous story about
Thales, in which the philosopher fell into a well while stargazing. It was,
for Tertullian, a metaphor for those "who persist in applying their studies
to a vain purpose . . . [and] indulge a stupid curiosity of natural objects."

But some Christian contemporaries of Hypatia did share her thirst
for understanding and love of classical knowledge. One was Augustine of
Hippo. By the time of his death, he had become a far more important figure
in the formation of the early church than Tertullian or anyone else, save
the apostle Paul. He was probably the most important thinker and writer
in the first two millennia of the church's history. And he espoused a very
different strain of thought than Tertullian did—one in which the knowledge and learning of the Greeks was to be cherished and not discarded.

Augustine came from modern-day Algeria, and he spent most of his
life on the North African frontier of the Roman Empire, first as a teacher of
rhetoric in Carthage and later as the bishop of Hippo. As a young man, he
always seemed to be in search of something. He turned to Manichaeism,
losing himself in the mystical teachings of the Iranian prophet Mani, who
borrowed from Christianity and Buddhism. For a short time, Augustine
became a Neoplatonist as Hypatia had been, absorbing himself in Greek
classics that would leave a lasting impression on his young mind. Eventually, he converted to Christianity, the religion of his mother.

Augustine's early writings, those penned most recently after his conversion, were closer to the inflexibility of theologians like Tertullian. But
as he matured in his faith, his writings became more open to the classical
teachings he had been exposed to in his youth. By the time he wrote one of
his most famous works, *Literal Commentary on Genesis*, he was urging his
fellow Christians back to the study of the natural world:

Even a non-Christian knows something about the earth, the heavens,
and the other elements of this world, about the motion and orbit of the
stars and even their size and relative positions, about the predictable
eclipses of the sun and moon, the cycles of the years and seasons, about

the kinds of animals, shrubs, stones, and so forth, and this knowledge he holds to as being certain from reason and experience. Now it is a disgraceful and dangerous thing for an infidel to hear a Christian talking nonsense on these topics; and we should take all means to prevent such an embarrassing situation, in which people show up vast ignorance in a Christian and laugh it to scorn. . . . If those who are called philosophers, especially the Platonists, have said things which are indeed true and are well accommodated to our faith, they should not be feared; rather, what they have said should be taken from them as from unjust possessors and converted to our use.

Augustine was intrigued by the mysteries of nature and wrote prolifically about them. He was a keen observer of plants and trees, which he looked at with the eye of a natural philosopher. He noted the medicinal qualities of hellebore, and he observed that hyssop could be used as a respiratory purgative. He noticed the seasonal growth patterns of plants, wondering why trees shed their leaves with the season, and why the leaves later grew back. He hinted at an understanding of osmosis that went beyond what most people understood for the next millennium.

Augustine also turned his inquisitive eye toward the animal kingdom. He accepted Aristotle's notion of spontaneous generation, even speculating that Noah had no need to take animals "which are born without the union of sex from inanimate things" aboard the ark. Spontaneous generation was all part of God's grand plan for the Earth.

Augustine's writings on the subject—and, thus, Aristotle's—would resonate in the Christian world for more than a thousand years. The theory even found its way into the dietary proscriptions of the church. Beginning in the twelfth century, Christians were allowed to eat geese on Fridays, when meat was forbidden. The policy was a result of English naturalist Alexander Neckam's "discovery" that geese could be born spontaneously from a mixture of pine resin and sea salt, which led to the widespread belief that they were fish. Even as late as 1623, William Shakespeare would write, in *Antony and Cleopatra*, "Your serpent of Egypt is bred now of your mud, by the action of your Sun"—a reference to the belief held by ancient

Egyptians that crocodiles were spontaneously generated by the action of sunlight on mud, which Herodotus had noted in his writings about the Nile. But Shakespeare wasn't just writing about what the ancient Egyptians believed. He was referencing what his seventeenth-century contemporaries, among them the most learned natural philosophers, believed. Some forty years after *Antony and Cleopatra* appeared, the most sophisticated scientific organization in the world, Britain's Royal Society, would hold meetings devoted to the ways serpents might be born from dust.

By Shakespeare's time, Europe was emerging from centuries of cultural stagnation, and a new spirit of learning and discovery was starting to take hold—an age of reason. With the dawn of the Renaissance, the ancient learning of men like Aristotle and Augustine would be reabsorbed, reexamined and, most important, questioned. Natural philosophers like Copernicus and Galileo would look to the stars and see a very different universe from the one that human beings had always known. Another would cast his inquiring gaze in a different direction and question the very idea of how life begins.

PROVANDO E RIPROVANDO

Do you see this egg? With it you can overthrow all the schools of the-
ology, all the churches of the earth.

—DENIS DIDEROT, *D'Alembert's Dream*, 1769

N THE WINTER OF 1662, three priests made their way through the streets of the Tuscan city of Pisa. The distinctive loud clicking of their shoes on the cobblestone avenues gave away their religious affiliation. More vulgar peasants would have called them *Zoccalanti*, after the heavy wooden sandals for which their order was known. The more reverential knew them as Franciscans, monks of the order of Saint Francis.

They were on their way to Pisa, winter home of Ferdinando II, grand duke of Tuscany. Ferdinando typically spent most of his time in the capital, Florence, but winters there tended to be wet and cold by Italian standards. Sometimes it would even snow. The grand duke was not fond of the cold.

Ferdinando had been a handsome prince in his youth. But the man who greeted the priests was middle-aged and fat, with puffy circles around his eyes. By then, Ferdinando had started wearing a mustache that curved upward at the wings, as if someone had painted a smile on his face. He looked a bit like a clown. Standing at Ferdinando's side was a young man named Francesco Redi, the grand duke's personal physician, confidant, and right-hand man in matters of science.

The grand duke had a reputation for lavishing generosity on those who could furnish him with anything he might consider a scientific wonder. The Franciscans had just returned from the Orient, and they were carrying many such gifts for the duke. They were particularly excited about some tiny black stones they had brought back from the region around the Ganges River that had been extracted from the head of a snake that the Portuguese called a cobra. The stones, the priests said, were wards against all forms of poison, whether delivered from the fangs of a serpent or the poisoned weapon of a man. One need only apply these stones to the wound and they would stick, like magnets, until they had absorbed all the poison into themselves. The stones could then be washed with freshly squeezed milk and the poison would be released, leaving them ready to be used once more.

Redi had seen such stones before. Their supposed supernatural properties were well known to those who practiced the healing arts. Their efficacy had been vouched for even by the Roman physician Galen, one of the most important figures in classical medicine. But Redi was not impressed. He was a skeptic at heart, active in the Florentine scientific society called the Accademia del Cimento ("Academy of Experiment"). The group's motto was *provando e riprovando*, "test and test again," and Redi embraced it fully. Over the years, Redi himself had acquired many such stones—some from true believers and others from mere charlatans. None proved to have any more protective power than the pebbles one could find in any field.

Soon, the most learned men on hand in Pisa, many of them schooled in medicine, had gathered to see the miraculous stones brought back from the Far East. The grand duke decided to put the Franciscans' stones to the test. He called the guards to find him some vipers, but it was winter and none could be found. A new test was devised, using a chicken. A needle was dipped, four fingers deep, into a deadly poison made from tobacco, and the chicken was stuck with it. The stones proved useless, and within a quarter of an hour the chicken was dead. The Franciscans were flabbergasted. Try again, they urged. More chickens were dispatched, one after another. The priests were never quite convinced, arguing that the chickens had died from some cause not readily apparent. Many years later, Redi wrote of the episode in a letter to the famous Jesuit natural philosopher Athanasius

Kircher: "Doubt often wants to grow at the foundation of truth," he wrote, "like a blooming sprout."

REDI HAD ARRIVED at Ferdinando's court just a little more than two years before the Franciscans. As the grand duke's physician, he was entitled to a suite in the Palazzo Pitti, the grand duke's marvelous palace in Florence. The palace itself was a kind of embodiment of the two very different ages between which Redi lived. On the outside, it was a Dark Ages fortress, with tall arches arranged in forbidding militaristic columns, built by Roman soldiers to protect against tides of barbaric invaders. Inside, however, it was filled with vibrant tapestries and daring art, the opulent trappings and symbols of the power of its owners, the Medicis, one of the most important families of the Renaissance.

The Medicis were bankers, and spectacularly successful. Their gold flowed through every corner of Europe. They were fond of showing off their wealth in their possessions, especially their collection of the most modern scientific contraptions. The Palazzo Pitti was filled with all manner of the latest technological gadgets: thermometers, astrolabes, and the world's first barometer. It even held the world's finest collection of telescopes, a lingering reminder of the famous astronomer who had once also walked the palace's halls: Galileo Galilei.

Though Galileo had died nearly two decades before Redi arrived at court, Redi would have been impressed by those signs of his famous predecessor. Galileo refused to accept the world as it was told to him. He sought to explain the world through observation. He embodied—as he still embodies—the struggle between reason and dogma. And Francesco Redi was firmly on the side of reason.

Like his father, Cosimo II, Grand Duke Ferdinando II was fascinated with all things relating to science. He even kept a collection of human specimens he considered scientific oddities, including a dwarf that roamed the palace and was said to have tusks in place of teeth. Some, like his devout wife, Vittoria della Rovere, thought him obsessed. He did little to conceal his hatred of her pious sermonizing. And there were other reasons for their marital discord. Gossip about the grand duke's love life was commonplace in the Florentine court. Some said Vittoria discovered him in bed with a man many believed to be his lover, Count Bruto della Molera. Ferdinando II's own mother, Archduchess Maria Maddalena of Austria, told a story of a cold winter day when she stopped in at his chambers. She had with her a list of Florentines with power and influence who were said to be sodomites. They should, she advised him, be burned at the stake. Her son looked over the list and told her it was incomplete. He added a name and handed back the paper. The name the grand duke had added was his own.

Redi thrived under the grand duke's patronage. He became active in the Accademia del Cimento, which had been started by the Medicis in order to keep Galileo's science alive after the great man's death. And always, Redi wrote prolifically on the natural sciences. Whenever possible, he stuck to the academy's motto of *provando e riprovando*. He often did so boldly. He once drank snake venom to prove that, although fatal when it enters the bloodstream, it is harmless when swallowed. Some of the greatest scientists of the time filled the ranks of the Accademia del Cimento, including the most important of Galileo's disciples. Yet of all its esteemed members, Francesco Redi would be the one most remembered by history.

In a world of European science dominated by Greek and Roman classical thinking, Redi was a new breed of natural philosopher. He was, in his younger years at least, skeptical of everything. Books were useful sources of facts. Knowledge was to be accumulated. But everything should be put to the test, from miraculous wards against poison to even the revered Aristotle's theories on the spontaneous generation of life.

In the tenth century ad, the Byzantine emperor Constantine VII commissioned a compendium of practical agricultural wisdom. Called the *Geoponica*, it became a sort of farmer's almanac for Europeans over the next six centuries. It was filled with useful knowledge like instructions for making wine or tips on breeding cattle. There was a lot about managing bees, something essential to productive agriculture. It even included a recipe for creating bees:

Build a house, ten cubits high, with all the sides of equal dimensions, with one door, and four windows, one on each side; put an ox into it, thirty months old, very fat and fleshy; let a number of young men kill him by beating him violently with clubs, so as to mangle both flesh and bones, but taking care not to shed any blood; let all the orifices, mouth, eyes, nose etc. be stopped up with clean and fine linen, impregnated with pitch; let a quantity of thyme be strewed under the reclining animal, and then let windows and doors be closed and covered with a thick coating of clay, to prevent the access of air or wind. Three weeks later let the house be opened, and let light and fresh air get access to it, except from the side from which the wind blows strongest. After eleven days you will find the house full of bees, hanging together in clusters, and nothing left of the ox but horns, bones and hair.

It read like a magical spell, but even as late as the Renaissance, it was still considered science. After all, it worked. The bee recipe had been around at least as long the Roman poet Virgil. There were multitudes of similar recipes covering all manner of creatures. They could even be found in the works of one of the greatest Renaissance scientists, the Flemish physician Johannes van Helmont.

Historians have long grappled with the term "Renaissance science" because the science of the period had two very distinct phases. The first was restorative, as men of learning sought to reinstate the work of Greek thinkers like Aristotle and Anaximander, which had been largely lost and forgotten to western Europe during the Middle Ages. The second phase was innovative, as they began thinking of new, original theories, and using the process of experiment to test these ideas themselves. Van Helmont had a foot in both phases of the Renaissance.

A native of Brussels, then part of the Spanish Netherlands, van Helmont had spent his formative years at university in Leuven. There, he had enthusiastically thrown himself into the works of Galen and Hippocrates, the great figures of classical physiology. But as his studies progressed, he began to become disenchanted with the classics, finding them empty and unconvincing. Many years later, he would write that while he once found such works "certain and incontrovertible," he eventually felt his years of study "were worthless." He gave away all the books he had acquired as a student. His one regret, he would often tell people, was that he had not burned them.

Van Helmont went on to become one of the most important natural philosophers of the early Renaissance. He made remarkable strides in the understanding of gases, becoming the first person to isolate carbon dioxide, which he called "gas sylvestre." It was he, in fact, who coined the term "gas."

As an experimentalist, van Helmont had few peers. Even fewer shared his commitment. He spent five years conducting his most famous experiment, carefully monitoring and recording the growth of a tree to prove his hypothesis that plants gain weight from water and air, not from soil as people had always believed. Thus he laid the groundwork for a future understanding of photosynthesis. He also delved into the nature of body fluids, which he called "ferments," such as stomach acids and semen. He connected them to the chemical reactions that caused changes in the body. This was a monumental—though underappreciated—step in humankind's understanding of how living things work. It anticipated the modern theory of enzymes, the large organic molecules that control all the chemical processes that sustain life. By the nineteenth century, many scientists would come to see these processes as the key element in what makes living things living.

Yet van Helmont's place in the pantheon of science was often contradictory. Even as he challenged the assumptions of the classics in order to forge his own conclusions, van Helmont remained obsessed with ancient mysticism and alchemy, including some of the most questionable ideas derived from Greek science. Van Helmont did not shy away from using the term "magic," which he embraced despite his deep Roman Catholic faith. He was fascinated by Aristotle's ideas about spontaneous generation and was considered one of the periods foremost authorities on the subject. He even created recipes for bringing various living creatures to life. His most famous was a recipe for creating mice. It involved mixing a sweaty shirt with grain in a barrel and waiting for the wheat to "tran-schangeth into mice."

Francesco Redi found van Helmont's recipes about as credible as the stones brought by the Franciscans, as he did all such recipes based on spontaneous generation. He decided to put the theory to the test. He chose to focus on the fly. As anyone could plainly see, flies were not born in a conventional sense; they simply emerged from all manner of filth. People could say with absolute certainty that there was no such thing as a fly egg for the simple reason that nobody had ever seen one. But Redi had an epiphany when reading an account of spontaneous generation found in Homer's *Iliad*. "What if it should turn out," he later recounted, "that all the grubs that you find in flesh are derived of the seeds of flies and not through the rotting flesh itself?"

During July, the time of year when flies seemed to be at their most numerous, Redi placed a snake, a fish, some small eels, and a piece of raw veal into four different flasks, tightly enclosing each after they had been filled. He then did the same thing with four more flasks. These, though, he left open, exposed to the air and any insects that might happen by. Just as Redi expected might happen, maggots appeared on the rotting flesh in the open containers, but not in those he had shut off from the air.

The results supported Redi's hypothesis, but he realized the experiment was not definitive. Anticipating his critics, he wondered whether the maggots failed to appear in the sealed jars because they needed air to survive. So he devised an even more ingenious experiment, using containers wrapped in gauze instead of sealed containers. Maggots did appear, but only on the outside of the gauze. For Redi, the only viable explanation was that flies had been drawn to the decaying meat, but, unable to penetrate the gauze, had laid their eggs on its surface.

School textbooks would one day remember it as the "Redi experiment." What made it such a seminal event in the history of science was not so much what Redi had proved or disproved. It was, rather, *how* he had done it: by developing a hypothesis and creating two very different sets of experimental conditions to test his theory. It was one of the earliest and finest examples of a controlled scientific experiment. *Provando e riprovando.*

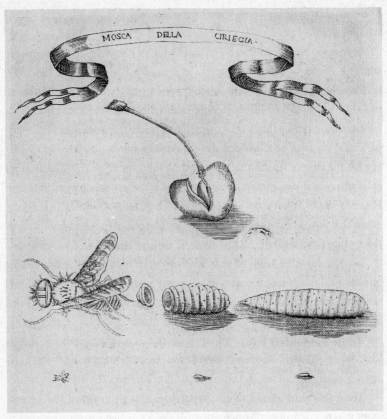

The life cycle of the fly, from *Experiments on the Generation of Insects*.

Soon, Redi was undertaking similar experiments with all kinds of insects. These formed the basis of his greatest work of science, *Experiments on the Generation of Insects*—for its time a masterpiece of careful observation and experiment. In it, Redi claimed to have disproved not only Aristotle's theory of spontaneous generation, but the very belief that nature, free of the hand of God, could have given rise to life. An artful writer, Redi summarized poetically the beliefs held by the classical Greek philosophers who thought that nature alone had given birth to life:

Many have believed that this beautiful part of the universe which we commonly call the Earth, on leaving the hands of the Eternal, began to clothe itself in a kind of green down, which gradually increasing in perfection and vigor, by the light of the sun and nourishment of the soil, became plants and trees, which afforded food to the animals that the earth subsequently produced of all kinds, from the elephant to the most minute and invisible animalcule.

To Redi, such a vision was incompatible with nature's laws. Echoing the words of the Dutch naturalist Jan Swammerdam, he wrote, "All life comes from an egg."

REDI'S CAREER IN THE SCIENCES would prove to be relatively short-lived. In May of 1670, Grand Duke Ferdinando II fell ill. The official cause was "apoplexy," a word then often used to describe what we would now call a stroke. Physicians attended him with the most sophisticated treatments they had at their disposal, applying hot irons to his forehead and smothering him with the flesh of dead pigeons. The treatments worked about as well as the stones said to ward against poison that the grand duke had once been given by the Franciscans. He died two days later.

The grand duke's only son, Cosimo III, assumed the throne. His father had wanted to give Cosimo a modern scientific education, but the duchess Vittoria would have nothing of it. The new grand duke was his mother's child in nearly every way. It was said that he never in his entire life smiled—a fact that his admirers took as a sign of his great religious devotion. His reign became remembered mostly for its oppressive laws against the city's Jewish population, which had grown to Italy's largest under his father's benevolent rule. He was also obsessed with chastity, establishing laws against making love near windows or doors, and even against women admitting young men who were not relatives into their houses. Homosexuals were beheaded. A biographer would later describe Cosimo as "a devotee to the point of bigotry; intolerant of all free thought; hated by his wife; his existence a round of visits to churches and convents."

Redi found his official position at court unchanged. He had long played the role of intermediary between Cosimo III and his father during their disputes, which were constant, and the new grand duke held Redi in some esteem. But Redi's career of scientific inquiry was no longer a viable option, and the younger Cosimo shuttered the Accademia del Cimento.

Redi instead threw himself into a new academy devoted to Tuscan literature, the Accademia della Crusca. He helped write the first Tuscan dictionary and authored several epic poems. He became far more famous, in his time, for his poetry than his science. His greatest work, *Bacco in Toscana*, is still considered a masterpiece of Italian literature. It revolved around a man's struggle to supplant the Roman god of wine. "So daring has that bold blasphemer grown, he now pretends to usurp my throne," laments the vengeful god Bacchus in Redi's epic poem.

Toward the end of his life, Redi's health began to fail from epilepsy. According to some accounts, he embraced Catholic mysticism, spending his days bathing in holy oil and spent a fortune on ribbons rubbed on the bones of Saint Ranieri, said to have miraculous healing powers. His own descriptions of his conditions seemed to imply that they were the result of nervous hypochondria.

Though Redi's great scientific work, *Experiments on the Generation of Insects*, was widely read, its true significance would have been lost to most people of the era. Redi had employed the tool of experimentation to ask whether life could truly come from nonlife. To him, it could not. And he was confident that he had proved the "fact" with an incontrovertible experiment. Yet few accepted that he had settled the question of spontaneous generation, because nobody could say definitively that they had seen a fly's egg. *Doubt often wants to grow at the foundation of truth.*

But meanwhile, far to the north, in Holland, an obscure Dutch haberdasher had acquired a copy of *Experiments on the Generation of Insects*. He was quite sure that what Redi was saying was correct, and not just because he believed in the infallibility of Redi's experimental methodology. He was quite sure because he had actually seen the egg.

THE EYE OF A GNAT

Suns are extinguished or become corrupted, planets perish and scatter across the wastes of the sky; other suns are kindled, new planets formed to make their revolutions or describe new orbits, and man, an infinitely minute part of a globe which itself is only an imperceptible point in the immense whole, believes that the universe is made for himself.

—PAUL-HENRI THIRY, Baron d'Holbach,
Le système de la nature, 1770

IT IS 1664. Late in the summer. A boat is being steered through a lake just outside the Dutch town of Delft. It is being guided by a man in his early forties, with a small hint of a mustache, about 2 inches long, that looks like it was drawn by pencil. He wears a light-brown wig that settles on his shoulders, typical of the Dutch middle class to which he belongs.

The lake is called Berkelse Mere. It is tiny, barely a lake, and marshy. In places, it descends into bog. Its depths are uneven and tricky to maneuver. The fishermen are used to it, for the fish are abundant in Berkelse Mere, and said to be uncommonly delicious. But this man is a stranger to the waters. He is a townsman from Delft, a merchant who deals in cloth. His name is Antonie van Leeuwenhoek, and he is looking for something in the lake.

There is another thing that Berkelse Mere is known for, although some think it may be related to the wealth of fish. In the winter, its water appears quite normal. It is exceptionally clear, in fact. But by early summer, the water in Berkelse Mere begins to take on a milky-white hue. Eventually, it fills with puffy green masses that float like clouds beneath its surface. The locals say the masses spring from the heavy dew that forms at this time of year, which they call honeydew. Van Leeuwenhoek isn't so sure. And he thinks he may know how to solve the mystery once and for all.

He steers the boat over to one of the floating green clumps, where he produces a glass vial. He scoops up some of the greenish water, taking it with him on the two-hour carriage ride back to his row house in town, where he lives with his wife and stepdaughter. He isn't sure what he'll find in the water. He has no inkling that the only world humankind has ever known is about to get infinitely bigger and that everything people think about the nature of life is about to be turned on its head.

He puts the specimen aside. For the rest of the day, he resumes his life as a modest draper who lives in a modest house in a modest town in Holland. Perhaps he attends to his business. Perhaps he spends time with his daughter, Maria, whom he cherishes, all his other children having died in infancy.

The next day, he returns his attention to the lake-water sample. With tweezers, he extracts a long, greenish strand, about the size of a human hair, from a droplet, ever so carefully. His mother was from an old family of brewers, and the strand reminds him of a copper *worm*, the coil used to cool beer and ale during the brewing process. He sets it within a strange contraption that he, himself, built. It is a metal plate, about 10 inches long, attached to a metal clamp that looks like something a carpenter would use to mount a device on a workbench. He uses this clamp as a pedestal for drawing things up to the center of his metal plate, where he has drilled a hole to hold a tiny piece of glass that he has ground into a lens. The device is called a microscope, and the haberdasher from Delft has constructed one that enables him to see things smaller than anyone else on Earth can see, or has ever seen.

Van Leeuwenhoek places a drop of water onto his device and carefully looks into the lens. He sees something. It appears to be a tiny white oval. What's more, it has what look like legs near what he imagines to be a head. And on the opposite side of the oval are things that look like fins. It is, he thinks to himself, a thousand times smaller than the smallest insect he has ever seen. When he sees it suddenly move, very quickly, darting like an eel through water, he is quite certain that it is alive.

FOR EVERY SINGLE human being on Earth, there are a billion trillion microbes. They thrive nearly everywhere—buried in rock 2,000 feet belowground, at three times that depth beneath the sea, and even within our bodies. A human being plays host to ten times as many microbial cells as human cells. Yet throughout most of history, humans were unaware of these ubiquitous life-forms with which we share the Earth. We wandered about, blind to the teeming jungle of tiny life-forms that we always mistook for a barren desert. The microscopic creatures that van Leeuwenhoek observed in his drop of water from Berkelse Mere were the first glimmers indicating that the world was *much* more densely populated by life than human beings had ever imagined.

It was fitting that the world of bacteria, protozoa, and countless other microbes was discovered during the seventeenth century. That century saw the world and human beings' understanding of it begin to expand in unprecedented ways, and the year of van Leeuwenhoek's birth, 1632, was important for a couple of very contradictory reasons.

On the one hand, 1632 marked the halfway point of the deadliest war Europe had ever known, or ever would know until the twentieth century. It was the war between Catholics and Protestants known, at the time, simply as the Great War. Historians would later give it the name by which it is usually remembered, the Thirty Years' War. The countryside of much of central Europe had been transformed into a vast battlefield reminiscent of the macabre paintings of van Leeuwenhoek's fellow Dutchman, Hieronymus Bosch. Whole towns and cities were wiped off the map. Even deadlier than the armies were the multitudes of diseases that followed in their wake. The population was beset by "head disease" and "Hungarian disease." Typhus, bubonic plague, dysentery, and scurvy all took their toll. Amid the chaos, fanatics engaged in mass slaughter and religious pogroms. Some fifty thousand women and men were accused of witchcraft and hanged, drowned, burned alive, or impaled on stakes.

But the year 1632 was also a year that foretold hope in Europe, the dawn of a new age of science and reason that would come to be known as the Enlightenment. If science had begun to trickle back into the intellectual landscape of Europe during the Renaissance, during the Enlightenment it would gush like a deluge. And an astounding number of the most important figures of the Enlightenment were born in 1632. One was the Englishman John Locke, whose concept of the natural rights of all people, as opposed to the absolute power of the monarch, would inspire Enlightenment thinkers such as Voltaire and Jean-Jacques Rousseau, and underpin democratic revolutions in the Americas and France. Another was the Dutch Jewish philosopher Baruch Spinoza, who turned to rationality to explore spirituality, viewing God not as the creator of nature, but as nature itself. He would leave such a mark on the world of philosophy that the German philosopher Georg Wilhelm Friedrich Hegel would one day remark, "You are either a Spinozist or not a philosopher at all."

Holland would become one of the centers of the Enlightenment. Delft, its capital at the time, would see two great figures born that year, van Leeuwenhoek and the painter Jan Vermeer, who would go on to create the masterpiece *The Girl with the Pearl Earring*. Vermeer's revolutionary use of color and light would carve out a place for him among the most influential artists in history. The houses where he and van Leeuwenhoek were born were just a short walk from each other.

VAN LEEUWENHOEK'S FATHER, Philips, was a basket maker who had married into a family of brewers, which was a respectable trade in seventeenth-century Holland. Philips had married above his station, which wasn't uncommon in the Netherlands. While the lives of most Europeans were mapped out by rigid lines of status and privilege, such distinctions were beginning to fall apart in Holland. The Enlightenment was opening the European world up to a much broader class of people. The Dutch began to believe that a man should be able to succeed because of his abilities, not the circumstances of his birth. Dutch women were also winning rights far beyond what women in most of Europe enjoyed or could even imagine. A gentlewoman could speak her mind freely and walk the streets unchaperoned without raising eyebrows. For the first time anywhere, wife beating became a crime.

Tiny Holland was rapidly becoming the center of European trade, with more ships than Spain, England, Portugal, France, and Austria combined. The Dutch had become the middlemen of Europe, moving goods between faraway colonies—those of their former enemies, as well as their own in places like the island of Java in modern-day Indonesia and New Amsterdam, situated on the island of Manhattan and destined to become the American city of New York. The Dutch began to talk of their *Gouden Eeuw*, their "Golden Age."

Along with Holland's newfound wealth came unprecedented freedoms, which helped make it a center of scientific progress in the Europe that emerged from the Thirty Years' War. Those freedoms extended even to religion, where the Calvinist Dutch believed in the separation between church and state. Communities of Jews, Lutherans, and even their recent enemies, Catholics, commingled and flourished in seventeenth-century Holland. At a time when religious strife so dominated the rest of Europe, a Dutchman like the painter Jan Vermeer faced few obstacles converting to Catholicism, while elsewhere such a conversion might have forced him to flee his nation or, at the least, might have hindered his career. Spinoza's writings, considered blasphemous at the time, earned the philosopher a writ of *cherem*—essentially an excommunication—from the Jewish temple to which he belonged, and he was reviled by the Calvinists. Yet his writing never landed him in prison, nor was he even in serious danger of prosecution. Nearly two centuries later, in England, the atheist poet Percy Bysshe Shelley would be tried for posting a single blasphemous handbill while he was a college student, and his atheism would eventually lead the British state to take his children away.

Immigrants from nearly everywhere were drawn to Holland's freedoms. Many, like Johannes van Helmont, were great scientific minds, and scientific theorizing flourished free of censorship from Rome. Dutch printing presses began to be filled with scientific treatises written in the Netherlands and abroad. Amsterdam became the first place where Galileo's banned work on the heavens, *Mechanics*, could be published. For many years, it was the only place.

The concept of a professional scientist did not yet exist. The very word "science," from the Latin word *scientia* ("knowledge"), was used sparsely, unlike the more common term "natural philosophy." Yet there was a class of great thinkers who we, in retrospect, classify as scientists. Most, like the physician Francesco Redi, had a day job. Van Helmont was likewise a physician, though he was often referred to as a philosopher. They were, by and large, men of a certain level of social status who could afford to devote time and money to studies that others usually saw as something of a hobby.

There was one thing all these natural philosophers had in common: education. Their ranks were filled by some of the most learned men of their eras. This was something van Leeuwenhoek did not share with his contemporaries. His father died when he was five. His mother and new stepfather sent him away for some perfunctory primary schooling. Latin and Greek were virtually mandatory for any fairly well educated person at the time; thus the playwright Ben Jonson could make light of his contemporary, William Shakespeare, for having "small Latin and little Greek." Van Leeuwenhoek had neither. He would go on to win credit for some of the most important scientific advances of the age, and to rub shoulders with the leaders of the world's most powerful countries. Yet he always seemed slightly out of place, insecure, and thin-skinned.

WHEN HE TURNED SIXTEEN, van Leeuwenhoek's stepfather passed away. His mother shipped him off yet again, this time to the port city of Amsterdam to learn a trade. The city was teeming with new arrivals, from abroad and from the countryside. The Netherlands was fast becoming an urban country, and Amsterdam, a great metropolis of its time. Van Leeuwenhoek took a job as an apprentice in a linen draper's shop, rising to the position of clerk and cashier, and learning the basics of the trade that would be his career for the rest of his life. It was probably then that he was first exposed to a simple microscope, the device for which his name would eventually become synonymous among learned people everywhere.

Simple lenses had been around since at least as far back as the first century AD. Emperor Nero's tutor Seneca the Younger noted that "letters, however small and indistinct, are seen enlarged and more clearly through a globe or glass filled with water." But it is hard to say when someone first realized that the lens-grinding process could be used to create what would qualify as a microscope. The Italian poet Giovanni di Bernardo Rucellai, who lived at the end of the fifteenth century and was a cousin to Pope Leo X, used a concave mirror to study bees. His observations became the basis for his most famous poem, *Le Api* ("The Bees").

Galileo had built one of the earliest compound microscopes, which he called his *occhiolino*, his "little eye." By stacking multiple lenses, he could produce a more powerful magnification. The term "microscope" was coined by Galileo's friend, the German botanist Giovanni Faber, who derived it from the Greek words *scopia* ("to see") and *micro* ("small"). Not long before Galileo, a pair of Dutch eyeglass makers named Hans Lippershey and Zacharias Janssen had both claimed to have invented the microscope. Both men also claimed to have invented the telescope. Fierce competitors who lived next door to each other, each insisted the other had stolen the idea for the devices. Neither claim ever had much merit. They were simply likely the first to try to *patent* such devices.

By van Leeuwenhoek's time, lens making was all the rage in Holland, and even Baruch Spinoza earned his living as a lens grinder. The most valuable lenses were used to make telescopes, since they could be used to navigate at sea and had important military applications. But there was a market for lenses that could see small things as well, particularly for those, like van Leeuwenhoek, in the textile trade. These were little more than what would later be known as magnifying glasses, but they could be used to accurately gauge the quality of cloth, and to closely examine the technique and skill of needlework.

In 1665, interest in microscopy was kindled by the publication of a marvelous book entitled *Micrographia*. Its author was an Englishman by the name of Robert Hooke, assistant to the famous Irish chemist and inventor Robert Boyle. As well as being a brilliant investigator of the natural world, Hooke was a gifted artist, and the book was filled with fabulous illustrations. These gave *Micrographia* a wide appeal, far beyond the narrow audience of those interested in a complex book of natural philosophy. Many of his subjects were mundane objects, but to a reader in the seventeenth century, the view from the lens of Hooke's microscope turned them into objects fantastic and magical.

Hooke's observations began with simple manufactured objects. Several were things a draper like van Leeuwenhoek might have chosen to observe. There were studies of the head of a needle and a piece of linen. Eventually, the book worked its way to more complex subjects, and Hooke turned his looking glass to plants, both commonplace plants like rosemary and exotic ones, like a plant brought from the East Indies called cow-itch. Finally came Hooke's most amazing and complex examinations, those of animals. He included everything from hair and fur and feathers to parts of insects

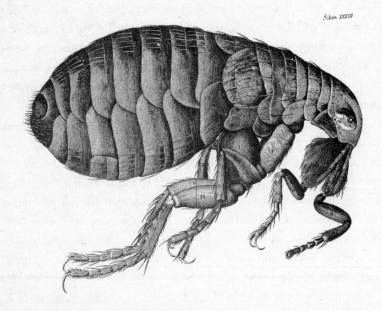

Hooke's drawing of a flea in *Micrographia*.

and other small organisms, like the eyes of the gray drone fly or the teeth of a snail.

One of the first illustrations in Hooke's *Micrographia* was a drawing of his microscope. It looked remarkably similar to an archetypical microscope design still used four centuries later, like a spyglass turned to the ground, with a small metal thimble upon which to place one's eye. Hooke included exhaustive instructions on how he constructed his microscope, even the methods he used to blow and grind his glass lenses. The instructions were so detailed that they filled much of the first half of the book.

BY THE TIME *Micrographia* appeared, van Leeuwenhoek was back in Delft, married and settled in a comfortable house in town. Soon he had built his own microscope based loosely on Hooke's design. It lacked the beauty of Hooke's tubular device, but van Leeuwenhoek was not completely blind to aesthetics. All of its component parts were made of copper or silver. When it came to the lens, van Leeuwenhoek made some changes to Hooke's design. Like all the most powerful microscopes of the time, Hooke's was a compound microscope. It had lenses stacked upon each other, each increasing the magnification of the last. Van Leeuwenhoek's microscope, in contrast, had but a single lens, yet he could see things that required a magnification five or six times greater than Hooke's provided.

Unlike Hooke, Van Leeuwenhoek was secretive about the methods he used to construct his lenses. He vowed never to share or even discuss them, and he never did, even when his secrecy threatened his own credibility. One modern observer, the artist David Hockney, has speculated that he used special techniques to increase the clarity of what he could see, including manipulating lighting or the background of his specimens. These could have been the same tricks used by many of the great Dutch painters of the period, who were masters of lighting and perspective. Hockney would also speculate that van Leeuwenhoek had been aided by something called a *camera obscura*—a simple box that, using light and mirrors, could project an unusually clear image that was much larger than the original, almost like a slide projector. Its design would eventually be used by the Lumière brothers, Auguste and Louis, as the basis of the first motion picture projector.

At least part of the reason van Leeuwenhoek's microscopes were so effective was that they were based on a *single*-lens design. The biggest problem with the compound microscopes of men like Hooke was that each additional lens obscured the clarity of what the observer could actually see—a phenomenon known as chromatic aberration. Van Leeuwenhoek's microscopes, which relied on one extremely powerful lens, didn't have that problem.

Van Leeuwenhoek was able to see things that no other human being on Earth had ever seen. He first turned his attention to the same mundane objects he found in Hooke's book, but saw details that Hooke had missed on the stingers and mouths of bees, and even in their eyes. He shared these discoveries with a few of his acquaintances, including Regnier de Graaf, a natural philosopher, accomplished physician, and inventor of one of the early prototypes of the hypodermic needle. De Graaf soon put van Leeuwenhoek in touch with Henry Oldenburg, an important natural philosopher in London. In the years to follow, van Leeuwenhoek would gain a reputation as the finest microscopist in the world. Oldenburg was the one who would make sure the rest of the world saw what van Leeuwenhoek had accomplished.

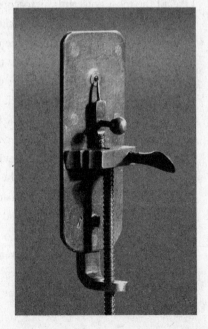

One of van Leeuwenhoek's microscopes.

A GERMAN FROM BREMEN, Henry Oldenburg's real name was Heinrich Old-enberg. He had first come to England as a diplomat but became a lifelong resident after marrying the daughter of an influential clergyman. He had a passion for the sciences. He was an early member of a small group of natural philosophers who met informally at London's Gresham College. They eventually adopted the name of the Philosophical Society of Oxford. In 1662, probably in response to the French court's support of a rival group of natural philosophers called the Montmor Academy, the Oxford group received a royal charter from King Charles II and became the Royal Society of London for Improving Natural Knowledge. Better known as simply the Royal Society, it rapidly became the best-known scientific body in the world, a distinction it would hold until the twentieth century.

The mathematician William Brouncker became its first president, and Robert Hooke was appointed curator of experiments. Oldenburg became the first secretary, but that appointment would be short-lived. In 1667, Old-enburg was arrested by British authorities and locked away in the Tower of London, suspected of espionage on the basis of a letter he had written to a friend in France, a fellow natural philosopher, in which he related events in the city. It had been a time of great xenophobia in London. A Dutch fleet was threatening England with invasion, and for the first time in their lives, Londoners could hear the guns of foreign ships off their shores. The city was also undergoing a severe outbreak of the bubonic plague, the last such outbreak in its history. Over the previous two years, the disease had claimed a hundred thousand lives. To make matters worse, a huge fire had burned down nearly 80 percent of the city the previous year. The process of rebuilding was under way, under the auspices of a brilliant young architect named Christopher Wren, another of the Enlightenment luminaries who happened to be born in 1632.

Oldenburg was released after the threat of Dutch invasion had passed. He wrote to his old friend Robert Boyle, whose children Oldenburg had once tutored, asking to be reinstated to the Royal Society and promising "faithful service to the nation to the very utmost of my abilities." Most welcomed him back, but there was a cloud over him that never quite went away. To many Englishmen, even some of those who knew him from the Royal Society, Oldenburg's loyalties remained in question. In later years, even Robert Hooke, suspicious of everyone and a staunch nationalist, questioned whether Oldenburg wasn't secretly in league with the French.

Yet Oldenburg became the organizational glue that enabled the Royal Society to establish itself as the world's premier repository of scientific thinking. His prolific correspondence with naturalists around the world made him a human nexus of Enlightenment knowledge. But the huge volume of letters he received from abroad also worried him. His arrest had made him cautious. He began asking his correspondents to address their letters to a "Mr. Grubendol," an anagram for "Oldenburg."

MICROGRAPHIA was the first major work published by the Royal Society. Originally, the project was to be completed by Christopher Wren, who was almost as accomplished a scientist as he was an architect. Citing a lack of time, he had passed the book on to Hooke. With funding from the Crown, the Royal Society also began publishing a periodical. Oldenburg became its first editor, and it didn't take long for the Philosophical Transactions of the Royal Society to establish itself as the world's scientific paper of record, a position it occupied for the next two hundred years.

Many of the *Philosophical Transactions'* early issues were devoted to microscopy. In 1673, the journal included a letter from Regnier de Graaf, the physician from Delft, who told of a "certain most ingenious person here named Leeuwenhoek" who had "devised microscopes which far surpass those we have hitherto seen." The claim was greeted skeptically. Nobody had ever heard of van Leeuwenhoek. An inquiry was made to the Dutch statesman and poet Constantijn Huygens, whose son Christiaan would go on to fame as an important mathematician and astronomer. Huygens wrote that van Leeuwenhoek was a man "unlearned both in sciences and languages, but of his own nature exceedingly curious and industrious."

At de Graaf's urging, van Leeuwenhoek dispatched his first letter to Oldenburg. It had all the charming reluctance and frankness that characterized his future correspondence. "I have no style, or, pen, wherewith to express my thoughts properly," he wrote. "Besides myself, in our town there be no philosophers who practice this art." He added a telling fact about his nature, one that he would never escape through all his future fame and success: "I do not gladly suffer contradiction or censure from others."

The letter included some observations of the stingers of bees and of lice, which could only have come from a microscope more powerful than Hooke's. There were also some simple sketches. Van Leeuwenhoek didn't have Hooke's artistic gift and never attempted a serious drawing. Later he had pictures drawn by local artists. Sometimes he showed them a simple sketch he had made upon which to base their own work, but he described these as no more than a few simple lines on paper. As far as anyone knows, he never let the artists simply look through his microscopes. To do so would have diminished his role as translator of the microscopic world, which for many years literally he alone had access to.

Van Leeuwenhoek's first communication was met with skepticism. That he was a mere haberdasher only increased the incredulity. Nonetheless, Oldenburg published an edited version in the *Philosophical Transactions*, adding a mildly sarcastic commentary of his own. Surely, Oldenburg wrote, they hadn't heard the last of this van Leeuwenhoek "who doubtless will proceed in making and imparting more Observations, the better to evince the goodness of these his glasses." Oldenburg was essentially inviting the Dutchman to prove he could see all that he claimed.

This, van Leeuwenhoek did. Over the next four decades, he dispatched some 560 letters filled with astounding scientific observations to the world's leading institutions and journals of natural philosophy. All of these were written in his colloquial style, meandering through mundane topics even as he made astounding scientific revelations. But van Leeuwenhoek never published a book or even what might generously be called a scientific paper. He probably never adopted a style more suited to publication because he couldn't read the foreign journals in which his work appeared. The only language he ever mastered was Dutch.

Most of van Leeuwenhoek's letters were directed to the Royal Society, and Henry Oldenburg became, to a large extent, his personal translator and editor. It is ironic that van Leeuwenhoek would owe so much of his initial fame to a German. Van Leeuwenhoek didn't care for Germans, and when speaking of them, he had a habit of turning to one side and exclaiming, "Oh, what a brute!" Until his death in 1677, Oldenburg dutifully edited all of van Leeuwenhoek's communications. Many of these were addressed to "Mr. Grubendol."

MOST PEOPLE in the seventeenth century had a hard time accepting the existence of things that could not be seen by the naked eye. Many of van Leeuwenhoek's claims were initially dismissed, even by the most learned of the time. Worse, for him, was the ridicule. Some of it came from the English satirist Jonathan Swift, who liked to poke fun at scientists. Van Leeuwenhoek's observation of parasites that fed on fleas inspired this parody by Swift:

> *The vermin only tease and pinch*
> *Their foes superior by an inch.*
> *So Naturalists observe, a flea*
> *Has smaller fleas that on him prey,*
> *And these have smaller still to bite 'em*
> *And so proceed ad infinitum.*

Yet nothing could have prepared van Leeuwenhoek for the incredulity that would greet his first great discovery, the microscopic world he examined in Berkelse Mere. It was a landmark moment in science. No other person had ever observed the tiny creatures that would later be understood to be protozoa and bacteria. Nor would anybody see them, without van Leeuwenhoek's personal assistance, for the rest of the century. He was the first human being to see a single-celled organism—a discovery for which he would forever be remembered as the father of microbiology. He called these tiny creatures "animalcules," or "little animals." He estimated that his tiny sample of lake water contained millions of them.

Van Leeuwenhoek was at first reluctant to share the discovery. More than a year passed before he finally described the animalcules in a letter to Oldenburg. Van Leeuwenhoek had guessed, correctly as it turned out, that he would not be believed. To most people in the late seventeenth century, the existence of creatures so small was beyond imagining. Many thought he was literally insane. Their suspicions only grew when van Leeuwenhoek would not provide them with a microscope capable of verifying what he had seen. Not until his later years, when his notoriety began to attract famous visitors, even royalty, did he give away some as gifts. Even then, the recipients complained that those they were given didn't work like the ones they had gazed through in van Leeuwenhoek's home.

If van Leeuwenhoek would not send his microscopes to them, then the Royal Society decided it would go to his microscopes. A delegation of several prominent British and Dutch clergymen arranged to travel to Delft. They eventually confirmed van Leeuwenhoek's claims. Within a few years, his reputation soared among the international scientific community. In 1680, the Dutch haberdasher became a full member in the Royal Society. He never attended a meeting, though, not even his own induction ceremony.

Four years after his observations of Berkelse Mere, van Leeuwenhoek made another monumental discovery. While examining his own saliva, he decided to look at the plaque between his teeth, which he described as "a little white matter, which is as thick as if it 'twere batter." There again, he discovered animalcules, little string-like creatures he described as looking like tiny eels. He judged he had seen a thousand in a bit of plaque "no bigger than a hundredth part of a sand-grain."

Soon, he was examining the dental plaque of almost everyone who would participate. In an old man who "never washed his mouth his entire life," he found plaque swarming with animalcules. Yet in the plaque of another old man, "whose teeth were uncommon foul," he found none. He guessed—probably correctly—that the absence of animalcules was because the man was so fond of wine and brandy. In a letter to the Royal Society that became one of his most famous, van Leeuwenhoek wrote "that all the people living in our United Netherlands are not as many as the living animals that I carry in my own mouth this very day." He wondered if the knowledge of their existence would prove too much for people to bear.

This was to be one of van Leeuwenhoek's greatest discoveries. Already the first to see simple one-celled organisms, he had just discovered the existence of bacteria, among the oldest forms of life on the planet and the source of so many diseases and infections. Of course, nobody then understood the true significance of what he had seen, or that it would, two hundred years later, contribute to a revolution in medicine.

At the time, his contemporaries were much more concerned with the question of where van Leeuwenhoek's animalcules came from. Spontaneous generation was seized upon as a possible method for microbes to propagate themselves, but van Leeuwenhoek suspected differently. Francesco Redi's *Experiments on the Generation of Insects* had been published in 1668, and the book had a strong impact on van Leeuwenhoek. He became convinced that Redi was correct, and that all life came from an egg. Some of his earliest letters contained thinly veiled attacks against spontaneous generation. "It was just as impossible," wrote van Leeuwenhoek in a 1686 letter, "for a louse or a flea to come into being without procreation as it is for a horse or an ox, or some such animal to be born from the decay and corruption of a dung heap."

Many of van Leeuwenhoek's investigations revolved around the subjects covered in Redi's book. What Redi had deduced through experiment, van Leeuwenhoek actually showed through the lenses of his microscopes. There, clear as day, anyone could see the eggs of flies or lice or fleas, or any of those creatures once thought eggless and parentless.

But in seemingly proving Redi correct, van Leeuwenhoek had settled one question, only to pose another that would prove far harder to answer. Tiny insects did indeed have eggs, but what of these far tinier animalcules? Not even he could claim to have seen their eggs.

To most people, the very idea of van Leeuwenhoek's animalcules fornicating seemed ludicrous. Spontaneous generation seemed a much more likely explanation, but van Leeuwenhoek was skeptical. He steadfastly asserted that these tiny beings reproduced the same way most creatures did. He even convinced himself that he could see them in the act of copulation.

As a natural philosopher, van Leeuwenhoek was cast from the mold of Aristotle. He was an observer, not an experimentalist like Redi. But as his successes stacked up, he became more comfortable with his own knowledge and confident in advancing his own theories. He decided to settle the question of the animalcules' reproduction by doing something he almost never did. He devised an experiment.

The experiment van Leeuwenhoek came up with was simple. He took a pair of glass tubes and filled them with rainwater and ground pepper, a mixture he always found teeming with microscopic life. He heated both, which he knew from experience would kill the animalcules. Then he used the flame to completely seal one of the glass tubes. Without air, he imagined, nothing would be able to survive in the sealed tube. Two days later, he examined the tubes. As expected, the microorganisms had reappeared in the open tube. But when he unsealed the second tube, he was surprised to find them present there as well. The simplest explanation was that the creatures in the sealed glass tube had spontaneously generated.

Van Leeuwenhoek never did bring himself to accept the easy explanation that the microbes he observed in the sealed tube had been spontaneously generated. Though he dutifully reported his results to the Royal Society, he was unsure what to make of them. For the most part, he simply moved on.

IN 1698, Peter the Great, tsar of Russia and one of the most powerful monarchs in the world, embarked on a tour of the Netherlands to inspect the military capabilities of his allies. A riverboat carried him through the canals from the Hague and into Delft, where he dispatched two of his adjuncts to van Leeuwenhoek's home to invite him to visit. Peter would have gone himself to van Leeuwenhoek, the men explained, but the tsar didn't like crowds. When they met on the tsar's boat, van Leeuwenhoek was delighted to find that Peter spoke perfect Dutch. He brought the monarch a gift, a microscope upon which had been fastened the tail of an eel so that he could see for himself the process of blood circulating. The tsar was delighted. Van Leeuwenhoek told acquaintances that he had found the encounter rather boring.

Van Leeuwenhoek had by then become one of the most famous men in the world. He had discovered hosts of microbial creatures. He was the first person to see spermatozoa in semen, and he was one of the first to see blood flow through capillaries, which he described in great detail. He had even described things as small as a single cell. In a 1692 essay on the state of microscopy, Hooke complained that the field had been "reduced almost to a single Votary, which is Mr. Leeuwenhoek; besides whom I hear of none that make any other Use of that Instrument except for Diversion and Pastime."

The accolades were never enough for the Dutchman. Even into his later years, van Leeuwenhoek complained to anyone who would listen about the ill treatment that had greeted his earliest and greatest discoveries. He kept working into old age and continued his correspondence with the Royal Society and others. But even those letters often betrayed a kind of bitterness that had long since passed being appropriate. He would frequently provide lists of "witnesses" to confirm even the most mundane observations, even though his reputation was, by then, long beyond reproach.

By 1723, van Leeuwenhoek was suffering from increasingly violent lung spasms that made it hard for him to breathe. He began writing about his condition in a series of letters to the Royal Society. Though blind by then in one eye, he accompanied these letters with microscopic investigations of the midsections of sheep and oxen. Physicians attributed his episodes to a bad heart, but van Leeuwenhoek thought their diagnosis wrong. It turned out the doctors were, indeed, mistaken. Van Leeuwenhoek had a rare condition called respiratory myoclonus, which would later come to be commonly known as Leeuwenhoek's disease.

He had a friend translate his last two letters to the Royal Society into Latin, one of the languages he had never mastered. The letters had a certain macabre character. Approaching ninety years of age, van Leeuwenhoek knew he was dying, yet he approached his end very clinically. One of his letters described a fit that lasted three days, "during which time my stomach and guts ceased to perform their office and motion, so that I was persuaded I stood at death's door."

Van Leeuwenhoek's condition gradually worsened. By August 1723, he was dead. He was buried in a cemetery in Delft, just yards from the grave of Hugo Grotius, the religious theorist whose ideas had formed the theological underpinning of the Methodist and Pentecostal movements, and one of the most important figures in the history of the Netherlands.

Van Leeuwenhoek left his most prized possession to the Royal Society: a beautiful black-lacquer box that housed twenty-six silver microscopes, upon which he had permanently affixed specimens, all arranged like the "cabinets of curiosities" that were so popular during the era. On receipt, a clerk at the Royal Society dutifully recorded the contents in almost poetic fashion: "The eye of a Gnat . . . Globules of Blood, from which its Redness proceeds . . . The Vessels in a leaf of Tea . . . The Organ of Sight of a Flie." The bequest was accompanied by a letter from a friend of van Leeuwenhoek asking that, upon receipt of the cabinet, the society send word to van Leeuwenhoek's daughter, Maria, "a spinster of excellent repute, who has preferred a single life to matrimony, in order that she might ever continue to serve her father." In 1739, Maria van Leeuwenhoek had a small monument erected to her father in the cemetery where he lay. Six years later, she was buried by his side. She had never married.

Antonie van Leeuwenhoek is remembered as the "father of" a host of scientific disciplines and subdisciplines—most important among them, microbiology.* That he began as merely a simple tradesman makes his accomplishments all the more extraordinary. After his work was done, the world became a much grander place, filled with an infinite array of microscopic life that no human being before him had ever guessed existed.

The implications for humankind's understanding of the origin of life were enormous. Van Leeuwenhoek was a deeply religious Calvinist. Life, for him, was the work of God's hand alone. Such beliefs were held by almost everyone at the time, and they did little to prejudice him against spontaneous generation. His advocacy of the observations of Francesco Redi is

---

* Nearly 400 years after his death, a poll taken by a news organization in the Netherlands named van Leeuwenhoek the fourth greatest figure in Dutch history, ahead of Rembrandt and van Gogh.

what led van Leeuwenhoek to try to shut the door on the theory of spontaneous generation. Yet van Leeuwenhoek's experiments opened up an entirely new and even more contentious debate that would consume some of Europe's greatest scientific minds until the late nineteenth century: the debate over the possibility that *microbial* life could arise spontaneously. For the next two hundred years, some of the greatest scientific minds in the world would grapple with the question of just what had taken place in the spontaneous-generation experiments of van Leeuwenhoek and others. Soon the question was going to take on important religious implications, finding itself at the center of a debate between those who believed in a creator God and those who saw life as something that nature was capable of creating on its own.

THE LABORATORY OF THE ATHEISTS

It's better to go along with the stories about gods than give in to what the natural philosophers call Fate. If there are gods there is some hope of appeasing them with a little worship; if not, we are ruled by something that no one can appease.

—EPICURUS, c. 300 BC

———————

O N JULY 1, 1766, in the small French town of Abbeville, a young man named Jean-François de la Barre, was taken from a prison cell and brought to a room where his legs were placed in a torture device known as a "Spanish boot." For the next hour, his jailors methodically crushed de la Barre's legs and feet before, according to some accounts, removing his tongue. When they were finished, they lifted him up and placed his body in the cart that would carry him to his place of execution. Around his neck, they hung a sign on which they had written "impious, sacrilegious and hateful blasphemer."

As a member of the nobility—a descendant, in fact, of Joseph-Antoine de la Barre, a former governor of New France, France's colonies in North America—Jean-François de la Barre was beheaded rather than hung. When they were finished, his executioners burned his body in a fire along with a copy of a banned book they had found in his room, the *Dictionnaire philosophique*, which had challenged the existence of miracles and mocked the literal truth of biblical stories. The ashes of both were then tossed into the nearby River Somme.

At the time of the execution, the author of the *Dictionnaire* was keeping a low profile at his estate near Geneva. He was a man of many paradoxes. He was, on the one hand, vain, narcissistic, and, where his own personal safety was concerned, cowardly. He had already been thrown into the Bastille once and had endured two exiles, once for merely writing a poem that suggested Adam and Eve had never taken a bath. He had no wish to repeat either experience. Yet he was, at the same time, a brazen and impassioned champion of reform, a self-styled enemy of injustice and ignorance. He had tried in vain to use his influence to stop de la Barre's execution. And that influence was considerable, for he was a writer whose skill, in his time, was matched by few. His sharp intellect had made him one of the most famous men in the world. He was certainly the most famous writer. His name was François-Marie Arouet. Most knew him only by his pen name, Voltaire.

Voltaire's *Dictionnaire* was really more of what would later be called an encyclopedia, a collection of essays on various wide-ranging subjects. The form had become extremely popular by that time. The book that established the genre was the *Dictionnaire historique et critique*, which had been written in Holland in 1697 by the man Voltaire had called "the greatest master of the art of reasoning that ever wrote," the French intellectual Pierre Bayle. Bayle's *Dictionnaire*, like Voltaire's, was extremely controversial. Bayle was a Huguenot, as the French Calvinists were known. He had fled the religious strife of France in the seventeenth century for the freedom of the Netherlands, where he was able to pursue his radical notions of religious tolerance and skepticism. Though Bayle always professed to have retained his Calvinist faith, his book implied that no reasonable person could believe the stories contained in the Bible. His critics called him godless. Some of his admirers said it too. Back in France, his writings led to the arrest of his father (a Calvinist minister) and of his brother. His brother would die in prison.

Bayle's *Dictionnaire* became the most widely read book of philosophy of the eighteenth century. Voltaire himself wrote a preface to a later edition. The book was also immensely influential in intellectual circles, helping to shape some of the greatest minds of the Enlightenment. Thomas Jefferson insisted that it be included among the first hundred books that would form the American Library of Congress. Publishers were soon flooded by similar books, each seemingly more radical than those that came before. Many of these works challenged religious conventions. Some had even begun to brazenly question the very existence of God, even though to do so was risky. It could even be deadly. In 1757, amid the reactionary climate that followed a crazed assassin's attempt on the life of King Louis XV, the French Parliament had passed a slew of repressive measures, including the death penalty for anyone who "composed and printed writings tending to attack religion."

Nobody understood the risks better than Voltaire. "It is dangerous to be right in matters where established men are wrong," he once wrote. There was a time in his life when he seemed to thrill in testing such dangers. He would even court them. But by the time of de la Barre's execution, Voltaire was seventy-one. He was still physically able and mentally sharp, which he liked to say was because he fasted periodically, drank thirty cups of coffee a day, and ignored the advice of doctors. Age, however, had made him cautious.

Voltaire had published his *Dictionnaire* anonymously in Geneva in 1764, using a publisher that specialized in dealing with forbidden books. There were many such publishers in that city, each with their own smuggling rings that specialized in slipping their wares into foreign countries. These same publishers also dealt in saucy books that passed for erotica at the time. The two genres sold extremely well, and their printers put both in the category of "philosophic books."

No one was fooled about the authorship of the *Dictionnaire philosophique*. Voltaire was renowned for not being able to keep a secret, especially when it involved a project in which he had invested so much of himself. The book had taken him twelve years to write. He considered it his life's work, a compendium of all the wisdom he had gained and a summation of his philosophy. Yet when it was eventually banned and copies were burned in town squares all over France, Voltaire simply shrugged his shoulders and pretended not to care. Worse things could happen to a writer.

Voltaire had taken care to couch many of the book's more controversial lines, especially those regarding Christianity, using a style of writing called "reportage," as if he were simply reporting the opinions of others. In truth, his private views on religion were often even harsher. In a letter to his lifelong confidant, Frederick the Great of Prussia, Voltaire had once called Christianity "the most ridiculous, the most absurd, and bloody religion that has ever infected the world." It was not that Voltaire did not believe in God; he simply did not believe in an *active* God. "Is it not the most absurd of all extravagances," he wrote in the *Dictionnaire*, "to imagine that the Infinite Supreme should, in favor of three or four hundred ants on this little heap of earth, derange the operation of the vast machinery that moves the universe?"

AFTER THE PUBLIC FURY surrounding the de la Barre affair had faded away, Voltaire began composing a series of pamphlets based on his essay on miracles. These came to the attention of a traveling schoolmaster who happened to be visiting Geneva that year, who took it upon himself to publish a response. In an earnest style that contrasted strongly with Voltaire's rhetorical flourish and bombastic sarcasm, the schoolmaster wrote that the world was indeed governed by laws that God had set down, but from time to time, God needed to intervene. "Miracles," he said, "are very intelligible and believable for the loyal Christian."

Nothing could drive Voltaire into a rage more easily than a critic, and he never let any criticism go unanswered. To Voltaire, it was as if the man had thrown down a gauntlet. "To hold a pen is to be at war," he would often say. Now he really did see himself at war. It was as if he had channeled the whole de la Barre affair into this one argument. His penmanship grew more precise, as it always did when he was angry. His letters grew more legible. A heated exchange of epistolary pamphlets began.

Such debates were common in the eighteenth century. It was customary for such exchanges to remain anonymous, but by the fourth exchange of letters, Voltaire had learned his antagonist's identity. He was an Englishman named John Turberville Needham, a Catholic priest. Needham seemed the embodiment of everything Voltaire detested: the church, sanctimony, superstition. Voltaire thought him an utter simpleton.

Needham could often be naïve. He tended to trust in the good nature of others, sometimes to a fault. But he was by no means simple. Needham was an accomplished natural philosopher, microscopist, and one of the finest experimentalists of his time. His scientific explorations into the generation of life had made him famous. His work featured prominently in all of the most important scientific journals of the time, and he had become the first Catholic priest ever admitted to the ranks of Britain's Royal Society. Above all else, he was known as one of the world's foremost authorities on the theory of spontaneous generation.

Gradually, Needham and Voltaire's argument meandered into the fields of natural philosophy. Until his death fourteen years later, Voltaire would turn his pen to the natural sciences more than he had at any point in his life. His argument with Needham over miracles became a debate about the nature of life itself, and how life comes to be. It was often colored by the deep religious tensions of the time. It became one of the first glimmerings of an argument between religion and science, and reason and faith, that would continue, in one form or another, for the next two and a half centuries.

Their argument also contained some surprises. Each man found himself playing a role to which he was not accustomed. Voltaire, who had once said that every thinking man should "hold the Christian sect in horror," ended up as the champion of faith and the belief in a supreme being. Needham, a Catholic priest who believed in miracles, unwittingly provided a scientific legacy that would underpin a new understanding of the world being propagated by atheists. Voltaire, one of history's greatest ironists, would die never quite understanding the twist. Needham would live to see what he had done, and it would haunt him.

JOHN NEEDHAM was an English Catholic who came of age at a time when it was dangerous to be a Catholic in England. In 1688, the country's last Catholic king, James II, was deposed by the country's parliament and replaced by the protestant William of Orange, the grandson of William the Silent from Delft, and one of the many heads of state who had at one time or another visited Antonie van Leeuwenhoek to see his microscopes. William of Orange's installment set in motion a series of Catholic uprisings, known as the Jacobite rebellions, each of which was violently put down.

Needham was a minor British aristocrat from an old family that had been split in two by religion. Needham's father, the head of the family's Catholic branch, worried about the direction the country was taking and decided to send his young son abroad to Douai, France, a few miles south of Lille, where a school had been established for English Catholics fleeing the violence in their home country. While officially a seminary, the school could compete with the better universities of the continent. Needham quickly became its brightest star, winning a reputation as a brilliant experimentalist and natural philosopher. Many of his professors even considered him their superior. He was eventually ordained, but Needham decided to devote his life to scientific inquiry. He chose the life of a secular priest, one who has forsaken the right to perform clerical services for a regular career. A series of teaching posts followed, including a professorship at the English university in Lisbon, although he abandoned the position after little more than a year. Needham was a frail man, deathly pale, with delicate, effeminate features. He told friends the hot Portuguese weather didn't agree with him.

Soon after his return to London, he turned his attentions to the field of microscopy, the branch of science he found most compelling. Within a year, he had made an important discovery that would shape the course of the rest of his life. Needham had been studying a batch of blighted wheat under his microscope when something in one of his samples caught his attention. There were fine, off-white fibers he had never seen before. He thought they might tell him something about the nature of the blight and decided to see what would happen if he placed them in water. To his surprise, the fibers soon teemed with microscopic creatures.

A year later, he returned to the same batch of wheat and repeated the experiment. Once again, the creatures—like van Leeuwenhoek, Needham called them "eels"—came to life yet again, the water having seemingly reanimated them from the dead. In 1745, Needham published his work in the *Philosophical Transactions*, but he took care not to draw any broad conclusions, simply reporting the facts as he had observed them. A year later, more of his observations of little "eels" that had seemingly emerged out of a simple paste of flour and water were published in the journal.

Needham's papers were translated and subsequently published in Paris, where they caught the eye of Georges-Louis Leclerc, Comte de Buffon, the director of King Louis XIV's botanical gardens, the Jardin du Roi. Buffon was a polymath who excelled in the wide array of fields he set his mind to. He was one of the most intuitively gifted mathematicians of his time. He came up with a solution to one of the earliest-known problems in the mathematical field of geometric probability by determining the mathematical odds that a needle dropped from a certain height would land within a demarcated set of lines. The problem became known as "Buffon's needle." Yet it was in the natural sciences that he would leave his greatest mark.

As a young man, Buffon showed little evidence of the genius that would one day be so apparent. At university, he was an average student at best. Soon after taking a degree in law, his murky involvement in a duel forced him into exile abroad for several years. But he returned to Paris at a fortuitous time. The country was about to undertake a massive overhaul of its navy, and somebody was needed to study the strength of different timbers used to construct ships. Buffon had made some important friends by then, and the task was entrusted to him. By the time he finished, he had so impressed the minister in charge that he was handed the prestigious position at the Jardin du Roi.

Buffon greatly expanded the size and mission of the Jardin, transforming what had once been merely a glorified medicinal garden into a world-class botanical collection, adding a museum and a zoo, and gathering together some of the country's greatest botanists. About the same time that Needham began his investigations on wheat, it fell upon Buffon, as director of the Jardin, to produce an inventory of everything the garden contained. Buffon embraced the task with relish. In his hands, the simple inventory became a project to write a dictionary in the mold of Bayle's, except that it would be devoted entirely to natural philosophy. And by "natural philosophy," Buffon quite literally intended it to encompass virtually everything then known to human beings regarding the living world.

This broad scope included a subject that puzzled him: the generation of life. It was too important a subject to overlook, but Buffon was uncomfortable with the prevailing theories of the time. When Buffon read about Needham's experiments, he thought the Englishman was on to something, even if Needham himself didn't quite understand yet how important it was. Needham, Buffon thought, was someone he could work with to tackle the mystery of life's origin.

IN THE SEVENTEENTH CENTURY, most natural philosophers believed that every type of living organism on the face of the Earth had always existed, from the very beginning of the Earth's creation. Every organism—every dog, every bird, every human being, and every worm—had been created by God in the form of something called "germs." These germs were like the seeds of plants, scattered at the dawn of creation by God over the face of the planet, like a gardener would scatter a future crop. Germs were tiny, far too small to be seen even with the aid of a microscope. And each such germ contained even tinier germs, the germs of every successive generation that any creature would ever spawn. They were all stacked inside each other, like Russian nesting dolls. The infinite nature of the theory was the one thing that people had a hard time coming to grips with, but one of the theory's most influential proponents, the French philosopher Nicolas Malebranche, would point out that it was no harder to believe in germs than in the life cycles of plants.* "One can say that in a single apple pit," he said, "there would be apple trees, apples, and the seeds of apples for infinite or almost infinite centuries."

Some believed that, in humans, germs were contained in male semen. Others saw them in the female's egg. In France, the theory was called "embodiment"; in England, "preformation" or "preexistence." It wasn't just

---

* Malebranche's concept anticipated, in ways he could not have understood, the idea of genetic inheritance. In a sense, the information for making potentially infinite generations of apple trees *is* contained in a single apple seed.

conjecture. Proponents of preformation could see the evidence all around them in the natural world. The transformation of caterpillars into butter-flies was taken as a sign of God's blueprint unfolding. The bulb of a tulip with its endless unfolding layers seemed a clue to the infinite layers of tulips that would spring forth, one after another. In the tiny eggs of frogs, microscopists thought they could see future generations of frogs waiting to be born. Those who believed in preformation were never short of evidence.

The theory was an old one, but it had gained traction in the late seven-teenth century as a response to the theories of one of the most important thinkers of all time, René Descartes. Descartes's great contribution to the natural sciences was to use his principles of deduction to understand a world whose workings, in his eyes, resembled that of a machine. Descartes carried this mechanical view of the world into what he hoped would be a great treatise on physiology. Yet the act of creation—the most important piece of the puzzle of life—continued to elude him until his final years. In the end, Descartes had settled on a theory that was purely physical, though the details were always sketchy. It was based on the mixing of sperm—something that, at the time, females were also commonly thought to pro-duce. This mixing then led to a kind of fermentation in the womb. "If one knew what all the parts of the semen of a certain species of animal are, in particular, for example, of man," he wrote in the posthumously published *De la formation de l'animal* in 1648, "one could deduce from this alone, by reasons entirely mathematical and certain, the whole figure and confor-mation." He compared the process to that of "a clock, made of a certain number of wheels, to indicate the hours."

According to preformation, God alone was responsible for the ulti-mate act of creation. Descartes's version was creation by matter alone. His concepts of mechanical laws governing nature or the heavens could be accepted without discarding prevailing religious dogma. And indeed they were, especially in France. What most people couldn't accept was his argument that human life owed itself merely to the workings of this vast machine. This seemingly small distinction marked a line that few dared cross. To do so was to invite the accusation of being a materialist—of believing in a world without any role for a grand creator—or even of being an atheist. French author Bernard de Fontenelle summed up the worries of many when he asked, "Do you say that beasts are machines just as watches are? Put a male dog-machine and a female dog-machine side by side, and eventually a third little machine will be the result, whereas two watches will lie side by side all their lives without ever producing a third watch."*

By the time Buffon sat down to write his own treatise of natural knowledge, the doctrine of preformation was still in vogue, even with most natural philosophers. But preformation didn't sit well with Buffon. He was a materialist who saw the world much as Descartes did. Everything in nature, including the origins of living things, should be explicable by comprehensible laws. Preformation, he was sure, was little more than conjecture. But though he felt that Descartes's version seemed closer to reality, he was aware the details were lacking. Then, Buffon was confronted by two discoveries that he thought could be clues to how living things were generated.

The first of these discoveries was made in 1741, near the Dutch seashore, where two little boys had spent a carefree morning roaming around the grounds of their father's estate. They had come upon some tiny creatures in an inland pond, little curiosities, which were quarter-inch-long green specks that appeared to be wading in the water. The boys put them in a jar and took them home to their tutor, a Swiss naturalist named Abraham Trembley. Whether what Trembley's charges had found were plants or animals was not clear at first. Whatever they were, they moved slowly—so slowly it was hard to tell whether they truly moved at all. But as the weeks passed, Trembley was quite sure that they did, in fact, move, if only an inch or two a day, and that they were indeed animals.

Trembley's first thought was that he had discovered a completely new

---

* As humankind's understanding of living things increased exponentially during the eighteenth century, a Cartesian perspective of living organisms became harder to deny—even to some preformationists. The Genevan naturalist Charles Bonnet would echo Descartes when he wrote that "even the tiniest fibril can be imagined as infinitely minute Machines with functions of their own. The whole Machine, the great Machine, therefore is the result of grouping a prodigious number of 'machinules' whose actions are concurrent or converge." Bonnet's understanding of the organizational nature of life—machines composed of ever-smaller machines—wouldn't be out of place in the twenty-first century. Yet Bonnet was also a staunch advocate of preformation.

species. This would turn out to be untrue, as these tiny animals had already been identified by van Leeuwenhoek. The name van Leeuwenhoek gave them was "polyps," though they would eventually be commonly known as hydras. They were strange creatures, to say the least. Beneath the lens of a microscope, they looked like a cross of a snail, an octopus, and a plant. As Trembley tried to learn more about the little creatures, he cut some of them in half. He was shocked to see that both halves continued to live. He wondered whether he was just witnessing residual movement akin to a severed lizard's tail. Then something amazing happened. Each half polyp gradu-

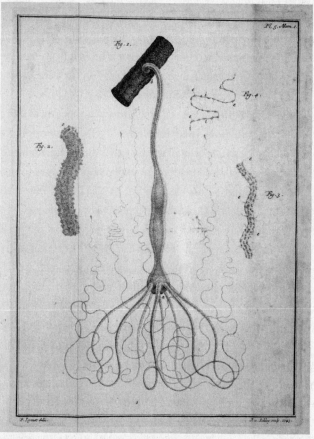

Trembley's hydra as depicted in his 1744 book, *Mémoires pour servir à l'histoire d'un genre de polypes d'eau douce.*

ally began to regenerate the portions of its body that it had lost. Amazingly, the two halves became two separate creatures.*

Trembley sent a summary of his results and a sample of the freshwater polyps to a well-known naturalist in Paris, René-Antoine Ferchault de Réaumur, an important skeptic of the doctrine of preformation who had written an influential paper about the regeneration of crayfish claws. Réaumur repeated Trembley's steps, cutting the odd specimens into sections. He, too, watched in wonder as the little creatures he had split formed into entirely new creatures. "I could hardly believe my eyes," he later wrote. "It is a fact that I am not accustomed to seeing after having seen it again hundreds and hundreds of times." When he presented a demonstration to the Paris Academy of Sciences later that year, the official report on the event compared it to "the story of the Phoenix that is reborn from its ashes," and asked witnesses to draw their own conclusions "on the generation of animals . . . and perhaps on even higher matters."

The conclusion Buffon drew was that life was not nearly as clear-cut as preformationists would have people believe. The freshwater polyp's ability to be split into two separate organisms did not, in his mind, fit the notion of preformation. He began looking for new explanations of how living things came about and found a possible solution in Needham's observations of "eels" springing to life in water, which seemed to revive the idea of spontaneous generation. Buffon decided more investigation was warranted. His book would not shy away from dealing with the all-important question of how life begins. If an answer to this question was not known, he was going to find the answer himself. He wrote to Needham in London and invited the Englishman to join him at his home in Paris for experiments.

In the spring of 1748, Needham arrived in Paris. Buffon's home was far more lavish than Needham could ever have imagined. His host was, by then, a fabulously wealthy man. Buffon's noble title had, in fact, been acquired through his outright purchase of the entire French village of Buffon, from which he took his name. He actually came from a family of civil

---

* In 1998, biologist Daniel Martinez discovered another amazing fact about Trembley's polyp: it does not age, because its stem cells have the capacity to regenerate, forever. Unlike most animals, the hydra is, in theory at least, immortal.

servants. He himself had been destined for such a career until he inherited a fortune from a childless uncle who had been tax collector for the entirety of Sicily during a period when the island had come under French rule. In the hands of a politically connected mathematical genius like Buffon, the fortune grew enormous.

The laboratory Buffon had set up for Needham filled an ornate drawing room, in the center of which was an intricately carved table well suited for fine dining, atop of which beautiful screens sat as dividers. This became the work space for Needham's microscopes, which he had brought with him at Buffon's request. Around the table, chairs were arranged so that Buffon's acquaintances could watch the two men at work.

Little else was frivolous about their collaboration. Buffon normally worked fourteen-hour days, even when his health began to fail in old age. While he compiled his grand inventory, a servant was paid a gold crown every day just to wake him at five in the morning. His motto was simple: "Do not waste time." He expected the same of everyone. Sometimes

Needham and Buffon examining a dog's testes.

Needham worked alone while Buffon attended to his responsibilities at the Jardin du Roi or conducted research for his massive book project. Sometimes the two men worked side by side. They set about dissecting the reproductive organs of dogs, rabbits, and rams. Under the lenses of Needham's microscopes, they examined the seminal fluids of all manner of creatures, even those of men, mapping out reproductive processes and searching for clues to a grand theory of how life began. They also searched for such clues in animalcules, where Needham's interest and experience were sharpest. They repeated his experiments on wheat and refined them, hoping to strengthen their case for spontaneous generation.

Needham began experimenting on mutton gravy, something he knew to be teeming with the kinds of microscopic life he wanted to understand. He took a dab of this gravy and enclosed it in a glass vial, which he sealed with a cork. He sealed it yet again with resin, to be absolutely sure it was airtight. He then heated the mixture. All of this care and preparation were to ensure that nothing could contaminate the experiment, that no living egg would be present in the tube, and that a microbe too small to see could not find its way into the mixture.

Needham was quite sure that no egg could survive in the miniature environment he created. Nothing could. He let weeks go by before he opened his sealed tubes. "My Phial swarmed with Life," he later wrote, "and microscopical Animals of most Dimensions, from some of the largest, to some of the least." The experiment was repeated with different substances replacing the gravy. Each time, the result was the same. First, Needham would observe tiny particles that he called "atoms." Day by day, these "atoms" grew and began to cling together until they became, in a couple of weeks, "the true microscopical Animals so often observed by Naturalists." It was, in many ways, a repeat of van Leeuwenhoek's experiment, with the same results that had so perplexed the Dutchman. For Needham and Buffon, it was clear that they had witnessed life emerge from nonlife, that they had witnessed evidence of spontaneous generation.

Needham published his work in the *Philosophical Transactions* in 1748, several years before Buffon would publish the results of their collaboration in his own book. This time, Needham went far beyond his first submissions. He had become bolder, more confident in the theories he had developed and perfected in the presence of Buffon's inquisitive mind. He didn't shy away from drawing conclusions that would make a mark on the world of natural philosophy. The spontaneous generation he witnessed, he argued, was not just the way *some* creatures reproduced. It was the method *every* creature used to reproduce. For, what were the spermatozoa he saw in his microscope if not little animalcules? He was convinced that these, and not the preformationists' "germs," were the true source of all life. He had discovered what he called a "vegetative force" that was the "one common Principle, the source of all, a kind of universal Semen." Needham added something else about his discovery, something that went beyond the purely scientific. He said that his "universal semen" was proof of an "an All-wise Being, All-powerful, and All-good, who gave to Nature its original force, and now presides over it." Not everyone would see it that way.

A YEAR AFTER Needham's paper appeared in the *Philosophical Transactions*, the first volumes of Buffon's simple "inventory" began to arrive in the bookstores of Paris. Released under the title *Natural History: General and Particular*, every available copy sold within a few weeks. The publishers rushed to keep up with demand. By the time Buffon was finished, *Natural History* comprised thirty-six volumes filled with gorgeous illustrations, including a plated illustration of Buffon and Needham at work in the laboratory.

In the second volume, Buffon unveiled his own grand theory of life. All organisms were composed of what Buffon called "organic molecules" that governed the life cycle from birth to death. For Buffon as it was for Descartes, life wasn't especially different from any other phenomenon found in nature. "Living and animation," he wrote, "instead of being a metaphysical degree of being, is a physical property of matter."

Buffon believed these molecules continued to exist even after death, traveling through the air until they joined with other organic molecules and reassembled themselves, forming entirely new creatures that could be different by grades from the original. This was how he accounted for the gradations within species that so many natural philosophers had begun to notice. His theory may not have been correct, but it was, for the time, awfully good conjecture, and a decent precursor to Charles Darwin's theory of natural selection. Even variations within the human race, Buffon believed, were a result of this phenomenon. Adam and Eve were Caucasians, while what he saw as lesser races were their descendants who had been made inferior by their weak environments.* As caustic as such views would sound a century or so later, they were almost progressive at a time when many thought of different races as altogether different species.

Buffon did something else in *Natural History* that would have historic implications. He used the word "reproduction" to describe the creation of offspring, a phenomenon previously referred to as "generation." The term "reproduction" had first been used by Réaumur to describe regeneration such as that found in Trembley's polyp, but Buffon used the word both in the regenerative sense *and* to describe the normal process of giving birth. While "generation" implied the blooming of a seed crafted by the hand of God, "reproduction" suggested replication, however inexact. Soon, more natural philosophers would begin to think of biology in terms of continuity, in terms of lines of descent.

VOLTAIRE FIRST READ *Natural History* in Geneva in 1767, not long after he had identified Needham as his foil on the subject of miracles. In his copy of Buffon's book, Voltaire scribbled angry notes in the margins. The word "chimera" appears over and over again, he noted. Where Buffon described Needham's work, Voltaire wrote, "Needham has seen, has imagined, has said only foolishness!"

Voltaire and Buffon had been friends in their youth, but Voltaire had

---

* Thomas Jefferson was one of many influenced by *Natural History*. The future president of the United States was bothered, though, by Buffon's characterization of the wildlife of the Americas—in Buffon's imagining, a land of swamps and bogs—as "inferior." Jefferson devoted the longest chapter of his only book, *Notes on the State of Virginia*, to rebutting Buffon's claim. In 1785, when Jefferson visited Paris, he took time to dine with Buffon at his home. He managed to sway the French naturalist, who removed any references to the inferiority of the American animals from subsequent printings of *Natural History*.

grown wary of Buffon by the time *Natural History* was released. The director of the Jardin du Roi was far too close to a man Voltaire considered
an enemy, the brilliant natural philosopher Pierre-Louis Maupertuis. As
Voltaire read through Buffon's book, he saw Maupertuis's influence everywhere, especially in Buffon's theory of life.

Voltaire's feud with Maupertuis was rather one-sided, and personal.
They, too, had been friends once. Both shared a passion for the work of
Isaac Newton, whose use of mathematics to describe natural phenomena
had revolutionized the study of physics. At the time, the French scientific
establishment was reluctant to embrace Isaac Newton, whose reputation
abroad was supplanting that of their own preeminent natural philosopher,
René Descartes. That Newton was an Englishman made it even worse.

Voltaire didn't share those prejudices. He spent his first exile in England
and was treated like a visiting dignitary. He received a pension from King
George II and was fêted by London's high society, including Jonathan Swift,
whose own work had influenced the younger French author. In letters to
friends at home, Voltaire had referred to France as "your nation" and noted
the "difference between their liberty and our slavery, their sensible tough-
ness and our superstition, the encouragement that all the arts receive at
London and the shameful oppression under which they languish at Paris."
He attended Newton's funeral at Westminster Abbey and set about learning everything he could about the Englishman whose use of mathematics
was transforming the study of physics.

No figure would ever have more influence on Voltaire's view of the natural world. Voltaire began slipping in the word "gravity" whenever he had
the chance, often merely to indicate his pleasure, similar to the way a modern writer would use the word to show seriousness. London, Voltaire said,
ran on "gravity." To Maupertuis, he wrote that he had become a convinced
Newtonian. "The more I glimpse of this philosophy, the more I admire
it," he said. "One finds at each step that the whole universe is arranged by
mathematical laws that are eternal and necessary."

A woman soon came between Voltaire and Maupertuis. Her name was
Émilie Le Tonnelier de Breteuil, marquise du Châtelet. By the time her
relationship with Voltaire started in earnest, du Châtelet was married, the

mother of three children, and on her way to becoming one of the most brilliant women of an era in which brilliant women were simply not recognized. She had largely stolen her education, sneaking her way into weekly seminars conducted by the country's most accomplished writers and scientists, and hosted by her father, the secretary to the king. As a little girl, du Châtelet devoured the lectures. Later, she became a regular at meetings of the French Academy of Sciences, where her young mind was quick to accept the brilliance of Newton, even while so many of the lecturers dismissed him. Voltaire had always ridiculed the notion of love, but anyone could see that he was smitten by the marquise. She had a mind that could compete with his own. It could even surpass it. He told friends he had met "his Newton." She said the two of them would be together forever.

In 1734, the fawning account Voltaire had written of his earlier exile in England was finally published in France. Voltaire had compared the two countries and found France wanting in nearly every respect. The treatise caused an uproar in Paris, and Voltaire fled the capital for a country estate owned by the marquise's husband. Soon the marquise herself joined him there. The house was set in the picturesque woodlands of Champagne, in countryside dotted with iron foundries built by the Romans. More important, the woods provided plenty of advance warning in case the authorities decided to arrest Voltaire, and it was close to the border in case they had to flee. The house was called Cirey.

Together, Voltaire and du Châtelet assembled a huge library of twenty-one thousand books on natural philosophy. The two set up a makeshift laboratory, where they would spend their days conducting experiments, most of which reportedly involved Voltaire's interests in the nature of fire. The room next door served as a bedroom for du Châtelet's husband, who would often come to visit. He knew of the affair. He didn't like it but stayed out of the way.

The two lovers wrote prolifically. Du Châtelet went to work translating Newton's great work *Principia mathematica* into French. Her translation would become the definitive French edition of one of the most important scientific books in history, and would still be used into the twenty-first century. Voltaire continued writing plays and commentary but also found time to write *Elements of Newton's Philosophy*, a book presenting the great thinker's theories in ways that were easy for regular people to comprehend. The book helped Newton win acceptance in France and popularized the story of the Englishman's "eureka" moment of the apple falling from a tree.

Eventually, Voltaire and du Châtelet drifted apart. Both took other lovers, and du Châtelet died in 1749, while giving birth to the child of the poet Jean Francois de Saint-Lambert. In a letter to Frederick, now King of Prussia, Voltaire wrote that du Châtelet was "a great man whose only fault was being a woman."

Throughout their affair, the arrogant Voltaire could never quite shake the fact that du Châtelet, not he, was the brilliant scientific mind. Du Châtelet likely wrote at least some of Voltaire's book on Newton, and may have written most of it. She understood Newton at a level Voltaire never could. What made it all worse was that du Châtelet had a man in her life who was her scientific equal. It was Maupertuis. She had known him since he tutored her as a girl, and had loved him once.

There was no evidence that Maupertuis ever returned du Châtelet's affections, but Voltaire could never shake his jealousy. In 1736, letters from Maupertuis began to arrive at Cirey, delivered from "the pole." Maupertuis was in Lapland, measuring a meridian to prove the shape of the Earth. At the time, many thought the Earth to be in the shape of an elongated globe, slightly longer about its axis, but Maupertuis was in the process of proving Newton's theory that just the opposite was true, that the Earth is, in fact, slightly flatter at the poles. Maupertuis returned a national hero and was named head of the French Academy of Sciences, the organization to which Voltaire would never so much as be offered a membership. Voltaire put his mind to seriously studying physics. He asked his friend Alexis Clairaut, a mathematician and astronomer with a reputation for simple straightforwardness, to assess his progress. Unimpressed, Clairaut told Voltaire to stick to literature.

In 1750, not long after du Châtelet's death, Voltaire went to Berlin, where Frederick had ascended to the throne. The king lavished his old friend with a 20,000-franc pension. But once again, Voltaire found himself overshadowed by Maupertuis, who a few years before had been named head of the Prussian Royal Academy of Sciences, even as he directed the equivalent organization in Paris. That was when Voltaire wrote his first of several satires pointed at Maupertuis, a story called *The Story of Doctor Akakia and the Native of St. Malo*.* Frederick had the book burned and even threw Voltaire in prison for a short time. Later, a book appeared making the accusation that almost everyone who visited the Prussian court knew to be true but none dared to say, that Frederick was a homosexual.† Voltaire was believed to be the author, and his friendship with Frederick was strained. They eventually reconciled in Voltaire's later years.

It was by way of Maupertuis that Voltaire first heard the name of John Needham. In 1752, Maupertuis published a series of letters detailing his thoughts on a range of scientific subjects. One of them delved into the subject of the origin of life, drawing on Buffon's *Natural History* and Needham's *Philosophical Transactions* paper. Maupertuis described Needham's experiment and the little eel that "looks to the eye like small fish" and "when allowed to dry and be without life for entire years, is always ready to be revived when one returns it to its element." But Maupertuis took Needham's observation to a place the priest had not intended. "Does not all of this," he continued, "plunge the mystery of generation back into an obscurity even more profound than that from which we had wanted to draw it?" The implication was clear, to Voltaire at least. For him, Maupertuis, like so many in the circle of French intellectuals in which they ran, was questioning the need for a creator.

Voltaire responded to Maupertuis's letters with disdain. He wrote another satire, *Séance memorable*, a parody of Berlin in which Maupertuis is depicted presiding over a lavish dinner that includes "a superb dish composed of paté of eels all one within the other and born suddenly from

---

* While in Berlin, Voltaire also found time to write a short story centered on a 120,000-foot-tall alien named Micromegas. The alien was a little like Jonathan Swift's Gulliver, except he traveled from planet to planet by "acquaintance with the laws of gravitation." He finds the Earth filled with "fools, knaves and miserable wretches." A philosopher tells him of "100,000 madmen of our species wearing hats, killing 100,000 other animals wearing turbans." Eponymously titled "Micromegas," it is considered by many to be the first work of science fiction ever written.

† As a teenager, Frederick tried unsuccessfully to flee the Prussian court with a man presumed to be his lover, Hans Hermann von Katte. His brutish father made Frederick watch von Katte's beheading. Frederick was then married off, though he and his wife saw each other only officially, and only once a year.

a mixture of diluted flour" and "fish that were formed immediately from grains of germinated wheat." At that point, Voltaire could still treat the science touted by Maupertuis as ridiculous. He would soon come to understand it as dangerous.

By THE TIME of his argument with Needham, Voltaire was an aging philosopher who no longer seemed quite so radical, at least in the radicalized world of French intellectuals. He had lived long enough that the world seemed to be passing him by. His support of enlightened monarchy seemed passé in light of the growing demands for a republican or even a democratic form of government. In the area of religion, where he had made such a mark as a provocateur, his views had begun to look tame. When he was young, those who questioned the tenets of religion were few and cautious. Their enemies saw little distinction among atheists, deists, or the many shades between. Neither, often, did the atheists or deists themselves. But things were changing. Men were becoming more daring, and those who called themselves atheists were becoming more brash.

Voltaire's views on religion, like his views on nearly everything else, were sometimes arbitrary and often contradictory. They were united in their hatred of superstition, and beyond that, little else. At times, his argument for God could appear utilitarian. He worried about whether morality could exist in a world devoid of a supreme being, a world in which good and evil were all relative. "If God did not exist," he wrote, "it would be necessary to invent him." Voltaire had a habit of quoting himself, a backhanded way of elevating his own importance. That quote was one of his favorites.

But there was a deeper reason for Voltaire's advocacy of the divine in the face of this now open disbelief: Voltaire really did see the natural world as proof of a divine intelligence. In the *Dictionnaire*, he had ridiculed the notion of an active God. But he did believe in an ultimate creator, what he called a "Supreme Infinite," responsible for creation, after which the world existed as it had always existed. Mountains stood where they stood because that is where God had placed them. They had never moved, as some natural philosophers, like Buffon, suggested. Nor had the seas or the forests. Fossils were not the remains of long-lost species, as some were beginning to suggest. They could tell us nothing about the world in which we lived or the creatures that inhabited it previously.

It was all too complicated to simply have happened by chance. These things had been carefully laid out. Nature had laws, but laws that fit the plan of a creator. They were an "intelligent design," as such a notion would one day be known. Voltaire's view of the world was much the same as that held by Newton, another deist. In the book on Newton that Voltaire had completed at Cirey, he had written, "If I examine on the one hand a man or a silkworm, and on the other a bird or a fish, I see them all formed from the beginning of things." The world may indeed be a clock, as Descartes suggested, but it had always been a clock. It had been, from the beginning, fully formed. Complete. "A watch," Voltaire said, "proves a watchmaker." If he detested organized religion, he detested atheism even more.

Slowly, as he grappled with the ideas of Maupertuis, Needham, and Buffon, Voltaire came to understand these ideas better than he thought their originators understood them themselves. Voltaire came to see the three men as a dangerous clique, one whose notions were leading men to materialism, after which atheism would inevitably follow. By the fifth letter of their exchange on miracles, he told Needham as much: "You had made small reputation for yourself among atheists by having made eels from flour, and from that you have concluded that if flour produces eels, all animals, starting with man, could have been born in approximately the same manner . . . from a lump of earth just as well as from a piece of paste." In subsequent letters, Voltaire's tone grew harsher. He began to call Needham "*l'Anguillard*"—the eelmonger—and an "Irish Jesuit." This last was to Voltaire one of the basest insults imaginable, as he had been schooled by Jesuits who he learned to detest, while he saw the Irish as hopelessly besotted by superstition. He also described Needham's microscope as "the laboratory of the atheists." Needham, like any proper Englishman, tried to ignore the name-calling. But the last charge, his complicity with atheism, was becoming harder to deny. It was all about to be proved true, in the pages of a book that would shock all of France.

By THE MID-EIGHTEENTH CENTURY, the axis around which Enlightenment Paris turned was a townhouse near its center that was called, alternately, the Hotel of the Philosophers, the Synagogue, or *La Boulangerie*, French for "the bakery." That nickname was a play on the name of one of the most radical men who used to frequent the house, the philosopher Nicolas Boulanger, who had written an audacious book suggesting that religions sprang from great disasters that occurred early in history. Men had turned to superstition simply to calm their fears of the natural world. Boulanger was a skeptic and a freethinker, and he fit right in at La Boulangerie.

The owner of the house called himself Paul-Henri Thiry, Baron d'Holbach. His real name was Paul Heinrich Dietrich. A German from just across the Rhine, he had spent five formative years among the radicals in Amsterdam and London before arriving in Paris, where he had taken a French name and a French wife. Like Buffon, one of the many frequent guests at La Boulangerie, d'Holbach had become fabulously rich through an inheritance from an uncle, to whom he also owed his aristocratic title. On the outskirts of the city, he owned a grand estate, complete with its own pastor, who d'Holbach kept long after he had stopped believing in God. Appearances and status were very important to him.

But d'Holbach's house in town was always at the center of his universe, as it was for so many of the intellectuals who were his frequent guests. Twice a week, he hosted lavish dinners for these friends. He was said to serve the best meals in town. At one time or another, he entertained David Hume, Adam Smith, and Benjamin Franklin. The brilliant young radical Denis Diderot was a close friend and a regular. So, for a time, was Jean-Jacques Rousseau. D'Holbach wrote several articles for Diderot's *Encyclopedia*, and probably served as the model for the character Wolmar in Rousseau's best-known novel *Julie, ou la nouvelle Héloïse*. His circle of associates was a radical one. Some may even have been as radical as d'Holbach. Few were as brazen.

During his early days in Paris, d'Holbach had been a deist with a worldview similar to Voltaire's. But by the 1760s, d'Holbach had become an ardent atheist, one who preached his disbelief with the fervor of the newly converted. Some said it was his close friend and collaborator Diderot who had influenced him, but there were plenty in his circle who shared this worldview. What set d'Holbach apart from these fellow travelers were the risks he was willing to take. In 1761, he published a book called *Christianity Unveiled*, an unapologetic assault on the idea of the existence of God. On the title page, he identified the author as the skeptic Nicolas Boulanger, his old friend, who had died before the book was penned. Using Boulanger's name was also an homage to the notorious house where d'Holbach's ideas had been spawned.

It was an ugly book. D'Holbach's writing style was plain, stilted, and hard to read. Diderot told him it lacked all art. But the book's sheer audacity gave it a certain power, and its lack of rhetorical flourish gave it a kind of widespread appeal, ironic since d'Holbach was an elitist snob, scornful of democracy and given to sneering at what he called the "imbecilic populace." The book found an audience, even though those merely caught with it in their possession were whipped, branded, and imprisoned by the French police. All too aware of the potential repercussions, D'Holbach took exhaustive precautions to keep his own involvement a secret, including making clandestine trips to London to meet with Marc-Michel Rey, a publisher in Amsterdam who had earned a reputation for orchestrating publication of the most dangerous books, including some by Voltaire.

More books on the theme of atheism were to follow, culminating in 1770 with the publication of d'Holbach's most influential work, *The System of Nature: or, the Laws of the Moral and Physical World*. It was an attempt to provide a clear-cut framework for the world, to answer existential questions without turning to a supernatural explanation. Like Karl Marx's *Communist Manifesto*, it was the kind of book that summed itself up in its first few lines: "Men will always deceive themselves by abandoning experience to follow imaginary systems. Man is the work of Nature: he exists in Nature: he is submitted to her laws." The soul does not exist, d'Holbach insisted. God was simply an anthropomorphization of things people did not understand.

In the second chapter, d'Holbach turned to the question of the origin of life. He touted Needham's experiments as proof that life could organize itself merely by physical processes, that nonliving matter could naturally become living. In a footnote, he invited readers to "see the observations microscopiques of M. Needham" and asked, "would the production of a man independently from the ordinary means be more marvelous than that of an insect from flour and water?"

The book was a scandal. Voltaire called it "a great moral sickness, a work of darkness." For Voltaire, at the root of that whole moral sickness was a singular idea, the idea that life could come from nonlife. To a friend he wrote that d'Holbach had "founded an entire system on a false experiment made by an Irish Jesuit who has been mistaken for a philosopher." Voltaire began referring to Needham's account of his experiments as "the story of the eels." It had become a creation myth for atheism. Adding to the list of his insults against Needham, Voltaire began accusing the latter of pretending to be Jesus Christ.

The System of Nature was perhaps even more distressing to Needham. "The world recoils in horror at the blasphemies hurled there against its Creator," he wrote. The use of his name was a personal insult. Yet even as he denounced the book, he was careful not to disavow or even waver in his own scientific observations and conclusions. They had, he said, simply been misinterpreted. His championship of spontaneous generation had made him famous. It had earned him recognition in the Royal Society and other pillars of natural philosophy. He could not let it go, even as it became clear that Voltaire had been prescient in his warning that Needham's microscope was "the laboratory of the atheists."

SOON, VOLTAIRE RECEIVED more ammunition for his feud with Needham, in the form of experimental evidence provided by an Italian university professor named Lazzaro Spallanzani. Spallanzani actually had much in common with Needham. Both were lay priests who had chosen a life in science and won fame as experimentalists. Spallanzani went on to become the first person to perform in vitro fertilization, which he accomplished using frog eggs. He also inseminated a poodle with semen he had taken from another dog, which was for a time thought to be the first example of artificial insemination, although Arab natural philosophers had, in fact, accomplished this with horses in the Middle Ages.

In 1776, Spallanzani went to work disproving Needham's conclusions on spontaneous generation. He repeated Needham's experiments, but more carefully, always searching for flaws in Needham's execution. Suspecting that air had leaked through Needham's corks, Spallanzani sealed his test tubes over a flame, fusing the melted glass until their contents were completely enclosed. He used all manner of substances: white beans, barley, maize, white beets, and egg yolk. Spallanzani's experiments were exhaustive, superior to Needham's in almost every way—a rare example of careful application of the scientific method for the time. By heating his flasks to different temperatures, Spallanzani showed that some animalcules could be destroyed only at extreme temperatures beyond what Needham had subjected them to. The two men then began their own public debate, which played out in their books and within the highest circles of natural philosophy.

Voltaire was quick to join Spallanzani's arguments to his own. Voltaire wrote Spallanzani almost sycophantic letters in the fawning tone usually reserved for his personal communications with royalty. In his published writings, he referred to Spallanzani as "the Italian savant," even as he continued to deride Needham as "the Irish Jesuit." Spallanzani was, in fact, more of a priest than Needham, having actually officiated at mass. The Italian had also been trained by Jesuits. None of that really mattered to Voltaire, though. Nor, really, did the experimental evidence. The crux of Voltaire's argument always remained the same: that positing a naturalistic explanation for how life came to be poses a challenge to the notion of a creator. The question of the origin of life, Voltaire understood, was important at a metaphysical level in a way that no other scientific question could ever be. Those were the stakes he was fighting for.

VOLTAIRE CONTINUED his relentless attacks on Needham, even into his last few years. They could still be found in his final book, *Dialogues d'Évhémère*, published in 1777. The following year, Voltaire died in Paris, where he had traveled to see the play *Irene* by the Englishman Samuel Johnson. It was his first trip to Paris since the de la Barre affair. The last words he wrote were, "I die adoring God, loving my friends, not hating my enemies, and detesting superstition." Church authorities in Paris denied Voltaire a Christian burial, but before their decision was issued, his admirers stealthily slipped his body back to Champagne to be buried in an abbey not far from Cirey, where he and Émilie du Châtelet had lived.

By the time of Voltaire's death, Needham had become head of the Imperial Academy of the Austrian Netherlands, as the territories of modern Belgium were then known. He received his appointment from the last Hapsburg ruler of the once-great Holy Roman Empire, Empress Maria Theresa, who had also appointed Spallanzani to head a university in Italy. Needham could never quite shake Voltaire's criticisms. He spent much of his public life refuting Voltaire's charges, sometimes also feebly reminding people that he was neither Irish nor a Jesuit. He even tried to fight back by trying his hand at a work of literature in the form of satire. He wrote a parody of a thinly disguised Voltaire who, Needham wrote, "misled our minds by the dictates of his heart." It fell far short in the art so carefully shaped by Voltaire. It was eventually published in the appendix of a book that Needham edited.

Needham's last attempt at experiment returned to the subject of miracles. An old Catholic folk belief held that the ringing of church bells could protect against lightning. French philosophers had taken to pointing out that, on the contrary, the lists of people killed by lightning annually always seemed to include an inordinate number who were ringing those same bells. Needham argued that church bells did, in fact, provide some form of miraculous protection. The work was mostly ignored. Where it was not, it was simply shrugged aside as an oddity.

In 1781, two years before Needham died, he penned a letter to a sympathetic French philosopher. As he explained the motivations that lay behind his experiments on spontaneous generation, he came very close to what might be characterized as regret. Preformation and the existence of germs, he believed, had never rested on firm scientific ground. These ideas simply weren't supported by the available evidence. Inevitably, natural philosophy would lead to new solutions to the problem of the origin of life. Needham's experiments were an attempt to provide a theory that could fill the void after preformation was proved wrong. He felt he had reconciled the mechanistic Newtonian world in a way that preserved a role for a creator God.

In fact, Needham had done the opposite. And in the coming century, the argument over the spontaneous generation of microscopic life would be at the center of an increasingly contentious debate that broke out of the insular realm of those devoted to natural philosophy and into society at large.

A VITAL FORCE

Oh! But it is all proved . . . read the book. It is impossible to contra-
dict anything in it. You understand, it is all science; it is not like those
books in which one says one thing and another the contrary, and
both may be wrong. Everything is proved.

—BENJAMIN DISRAELI, *Tancred*, 1847

T HE MANSION OF FYNE COURT was set deep in the sparsely populated
Quantock Hills of county Somerset, in southwest England. From
high points nearby, a person could take in a fine view of the surrounding
countryside. On a clear day, one could even see Glastonbury Tor, the cone-
shaped green hill that the ancient Britons once called *Ynys yr Afalon*, likely
the fabled "Isle of Avalon" of Arthurian legend. Yet the mansion itself was
set on oddly low ground, covered on three sides by dense forests in a valley
that was no more than a gash between the hills. No right-minded architect
would have chosen such a spot, a visitor once remarked, for it was as if "the
soil on which it had been built suddenly sunk one fine morning."

To the north of Somerset County, coal mining was uprooting the forests, fueling the steam engines of the first great industrial revolution. But
the Quantocks were largely spared. The woodlands were still filled with
unusually tall oaks and firs, the ground covered by wildflowers and ferns.
The poets William Wordsworth and Samuel Coleridge had spent summers
nearby. It was where Coleridge composed two of the most famous poems
in the English language: *Kubla Khan* and *The Rime of the Ancient Mariner*.

But in the unusually brutal winter of 1836, Fyne Court looked less like a preserve of England's rural past and more like a vision of an extraordinary future. At the top of each tree nearest the house, a long metal pole had been placed. From each of these, a third of a mile of copper wires were strung like Christmas lights. They crept from the bases of the trees toward the house, all converging on a single open window on the first floor, in the organ room. There, the wires made their way haphazardly past shelves stocked with vials of mysterious, multicolored liquids, until they reached a giant electric battery through which passed enough current to kill twenty men. The device bore a warning, written large and in Latin: *Noli Me Tangere*. Do not touch me.

The owner of the house, Andrew Crosse, didn't use the organ room much for entertaining. Crosse was a bit of a recluse, in his fifties and living the quiet life of a country aristocrat. His days were spent leisurely managing his lands and the tenant farmers who paid him rents, leaving him plenty of time to pursue his real interests. Crosse was known as a "scientist," a word that was only just replacing "natural philosopher" in the English vernacular. Electricity was his first passion, and he had by then earned a strong reputation for his work in the field. Interest in his work had even drawn a visit to Fyne Court from the president of the Royal Society.

Crosse's other interest was something that had intrigued him since he was a teenager: the natural formation of crystals. He thought the two subjects connected, that electricity might offer a clue to the question of how crystals were formed in nature. He began trying to form crystals by running electric currents through various pieces of stone, sometimes for weeks at a time. He thought that one day it would be possible to use electricity to create any kind of mineral, even gold or diamonds. The theory had gained him a measure of notoriety.

Crosse's latest experiment involved running an electric current into an airtight glass jar that contained a stone from Mount Vesuvius, the volcano that had destroyed the Roman cities of Herculaneum and Pompeii, and a solution of finely powdered flint and potassium carbonate. Every morning for the previous two weeks, he had donned the velvet smoking jacket he always wore in his lab and headed downstairs to check for results.

One morning, Crosse finally saw something that intrigued him.

Perched atop his rock were tiny white specks. Over the next four days, he dutifully returned each morning to find the specks growing ever larger. On the fourth day, he was stunned to see in place of one of the specks a tiny white insect situated just above the liquid. As the days passed, more appeared. They were, he scribbled excitedly in his journal, the "perfect insect." In the years to follow, Crosse's experiment would be cited as proof of spontaneous generation, as well as invoking comparisons to strange experiments conducted by the Italian Luigi Galvani, who had seemingly proved the existence of some kind of electrical life force. The experiment would also make Crosse, for a brief period of time, one of the more famous men in the British Empire, as well as one of the most vilified.

ONE MIGHT DESCRIBE the history of science as being shaped like an hourglass. Before the nineteenth century, science had always been largely a pastime practiced only by wealthy intellectuals, or those maintained by wealthy patrons. Its inaccessibility was not unlike that of the modern age, in which science is largely the exclusive purview of an educated elite we call scientists. But for a remarkable period during the nineteenth century, it became a pastime for nearly all walks of society. Particularly in the country at the center of the industrial revolution, Great Britain, science was discussed and debated in newspapers, at the dinner table, and even in working-class pubs and radical journals. A century earlier, few people would have had the opportunity to hear of experiments like Andrew Crosse's. Even fewer would have cared. By the middle of the nineteenth century, virtually everybody knew about the latest scientific trends, and no one seemed to be without an opinion.

This democratization of science was due largely to the advent of the steam-powered printing press. The business of printing was exploding. Paper prices tumbled to historic lows, and publishers saw mass publication as the wave of the future. Newspapers and periodicals popped up everywhere. Literacy, in turn, began reaching levels that had been unimaginable for most of human history.

As soon as Andrew Crosse's story found its way onto the pages of one of these new local newspapers, it didn't take long for it to spread nearly everywhere. Practically overnight, the quiet, unassuming man from Broomfield became a nineteenth-century version of a tabloid celebrity, the subject of drawing-room conversations throughout Britain and abroad. He became the scientist who created "life in a laboratory," and not just some invisible microbe. Crosse had supposedly proved the doctrine of spontaneous generation by creating a living, visible creature. His experiment was something that was beginning to pique the interest of wide segments of the public, the kind of thing the sensationalizing newspapermen of the day could use to sell papers.

Two books, in particular, were crucial in shaping the way Crosse's story was received. One was an immensely popular science book that appeared seven years after Crosse's experiment and held up his insects as proof of the creation of life purely through the laws of nature followed by a transformative process that led to more complex forms of life.

The other was a novel that had been published nearly twenty years before Crosse's experiment, one that whetted the public's thirst for new explanations of the mysteries of life. It had been conceived of on the shores of Lake Geneva, where a teenage girl had spent a rainy afternoon penning one of the most enduring works of fiction ever written.

IN THE SUMMER of 1816, eighteen-year-old Mary Wollstonecraft and her soon-to-be husband, the poet Percy Bysshe Shelley, traveled to Switzerland to visit the writer Lord Byron.* Shelley hoped to cultivate the more established Byron as a friend and mentor. While at Byron's villa on Lake

---

* In her preface to *Frankenstein*, Mary Wollstonecraft Shelley noted the extraordinary weather of the year 1816, since remembered as "the year without a summer." An abnormal cold had settled throughout the Northern Hemisphere during the summer months that year. Thirty inches of snow fell in Quebec City. In China, rice crops failed and water buffalo died off by the thousands. Heavy rains caused a cholera outbreak in the Ganges River in India that spread as far as Moscow. Summer frosts led to food riots in France and England. The freakish weather was probably caused by the most powerful volcanic eruption in recorded history. On the island of Sumbawa, in modern-day Indonesia, Mount Tambora had erupted with four times the force of Krakatoa and eight hundred times the force of the atomic bomb dropped on Hiroshima. Some seventy thousand people were killed. The atmospheric effects could be seen as far away as London, where for weeks the sunsets turned bright orange and purple. Many devout Christians saw these as signs of the coming of Armageddon.

Geneva, the group had planned a sightseeing expedition into the Swiss Alps, but heavy rains forced a change in their plans. Instead, they spent a legendary evening sharing German ghost stories. Byron challenged each in the group to write a story of their own. Only Mary and the English writer John William Polidori took up the challenge by writing novels. Published in 1819, Polidori's book, *The Vampyre*, was a classic in its own right, spawning a literary vampire craze that never really ended. But Wollstonecraft's novel had even greater and longer-lasting impact.

By the time of its publication, Mary Wollstonecraft had become Mary Shelley, and her book had a title: *Frankenstein: or, The Modern Prometheus*. The critics hated the book almost as much as the public adored it. What made Shelley's story unique—and so compelling—was the centrality of science. Her Dr. Frankenstein did not simply create a monster. He created a living creature from nonliving material, life from nonlife, just as Andrew Crosse was later supposed to have done. The book was, in a sense, a modern creation myth, one that reflected new ideas about the laws that govern life. And these ideas sprang from the science and skepticism of the time. It was a tale that could not have been written by the old German writers whose ghost stories were told that rainy day in Switzerland. Some have argued that it was the first science fiction novel.

In a preface to the 1831 edition of *Frankenstein*, Mary Shelley described the inspiration behind the story. Her plot stemmed from a conversation between Lord Byron and her husband about the experiments concerning spontaneous generation conducted by a "Dr. Darwin"—a description that has tended to confuse later readers. The Darwin she was referring to was Charles Darwin's grandfather, Erasmus Darwin, a larger-than-life character known as much for his brilliance as a scientist as for his eccentricities. He once famously turned down King George III's invitation to become royal physician. Corpulent and crippled by a childhood affliction with polio, Erasmus Darwin nonetheless fathered fourteen children by two wives and a governess, and scandalously prescribed sex as a cure for hypochondria.

In the sciences, Erasmus Darwin's beliefs were seen as radical. To some, they were even absurd. He was an early believer in transmutation, the theory that would one day come to be known as evolution. He held that species were subject to a gradual process of change that eventually led to new species and that was responsible for all the diverse organisms on the planet. The living world, Darwin believed, was molded by this process of transmutation, and the original source of life was the phenomenon of spontaneous generation. In his day, Erasmus Darwin was the most famous advocate of spontaneous generation in the English-speaking world. Both evolution and spontaneous generation often found a voice in his poetry. They were neatly summed up in his last and greatest work, the epic poem *The Temple of Nature*:

> *Hence without parent by spontaneous birth*
> *Rise the first specks of animated earth;*
> *From Nature's womb the plant or insect swims,*
> *And buds or breaths, with microscopic limbs . . .*

> *Organic Life beneath the shoreless waves*
> *Was born and nurs'd in Ocean's pearly caves;*
> *First forms minute, unseen by spheric glass*
> *Move on the mud, or pierce the watery mass;*
> *These, as successive generations bloom*
> *New powers acquire, and larger limbs assume;*
> *Whence countless groups of vegetation spring,*
> *And breathing realms of fin, and feet, and wing.*

Such notions weren't out of place in Mary Shelley's world. She was surrounded by freethinkers and religious skeptics. She traveled in the bohemian, avant-garde intellectual circles that were the English equivalent of La Boulangerie, its ranks filled by atheists who felt they had a particular stake in the question of life's origin. Her husband, Percy Bysshe Shelley, was probably the most famous British atheist of the time. Her father, the radical political philosopher William Goldwin, was perhaps the second most famous. In their eyes, like those of d'Holbach, God played no direct part in the creation of human beings. There had to be a naturalistic explanation for creation.

By the time of Andrew Crosse's insect-generating experiment, the connection with Shelley's mad scientist would have been an easy one for people to make. Both created life in a laboratory. Then there was the role of electricity in Crosse's experiment. The fact that Shelley had once attended one of Crosse's early lectures on electricity would, in later years, cause some to assume that he had been the inspiration for her Dr. Frankenstein. But such an association was highly unlikely, since at the time of the lecture, Crosse showed little interest in biology.

The analogy people drew between Crosse and Shelly's fictional scientist actually owed less to the book than to the wildly popular stage version of *Frankenstein* that appeared in 1823, which added electrical apparatuses as an embellishment. Shelley herself had left ambiguous the method by which Dr. Frankenstein revived his creature. At one point, she says he breathed "a spark of being into a lifeless thing." But the word "electricity" was never used. In the preface, however, Shelley did mention the influence on her text by the experiments of Luigi Galvani, a scientist whose own work had piqued interest in the possibility that human beings could come to harness the power to create life.

Galvani was an Italian professor of anatomy at the University of Bologna who had conducted an experiment that seemed at the time even more miraculous than Crosse's. Galvani had been studying a dissected frog leg, when he was startled to discover that the leg twitched whenever he touched it with his scissors. He suspected that this twitching had something to do with the electrical storm raging outside his laboratory. Later that year, the same thing happened when a crude electric generator was left on during a dissection. Galvani was by nature a cautious man and didn't easily jump to conclusions. "So easy it is to deceive oneself in experimenting, and to think that we have seen and found that which we wish to see and find," he once wrote.

Gradually, though, Galvani came around to the idea that he had hit upon a life force that he called "animal electricity." His experiment became rather infamous, mostly because of the showmanship of Galvani's nephew,

Giovanni Aldini. Aldini delighted in public demonstrations of his uncle's "animal electricity," and even took his show on the road to London. In 1802, he stimulated movement in a dead ox before an astounded audience that included King George III's wife, Queen Charlotte, and their son, the future King George IV. A year later, Aldini electrically animated the head of an executed criminal in front of some of London's most important physicians. He later recounted that "the jaw began to quiver . . . and the left eye actually opened." His uncle's discovery eventually became so renowned that it spawned the word "galvanize," meaning "to stimulate" or "to bring to life."

In a way, Galvani was right, although not in the way he thought or in ways anyone of his era would have been able to understand. Living cells are, in effect, miniature batteries, powered by the charge differential they maintain across their membranes, which is transformed into work by the molecular pumping of ions. In animals, the discharge of this electrical potential mediates the transmission of nerve signals, which in turn acti-

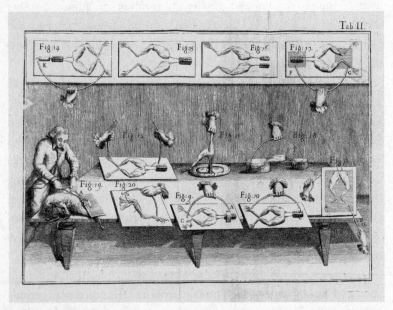

Galvani's frog leg regeneration.

vate the muscles. Electricity, in essence, keeps the heart pumping, operates the limbs, and creates the phenomenon of consciousness. But the idea that Galvani's famous experiment proved the existence of a special life force was eventually debunked by Alessandro Volta, whose name now graces our unit of electrical potential. In order to refute Galvani's experiment, Volta created the first example of an electric battery, now known as a voltaic cell. In his investigation into electricity's role as a life-giving force, Volta paved the way for a second industrial revolution—this one powered by electricity.

THE FIRST PERSON known to have written about electricity was the Greek playwright Aristophanes, who noticed that after amber is rubbed with a swath of fur, it exerts a pull on lightweight objects like feathers. Anaximander's mentor in Thales had observed an even more striking example in the way a piece of magnetite, commonly called lodestone, exerted a pull on anything made of iron. Thales attempted to describe the phenomenon, but as he did with most subjects that he found hard to explain, he turned to metaphysical explanations. The lodestone, he thought, must have a soul that was exerting power. Had Anaximander written about the phenomenon, he likely would have described it differently. But Thales's conclusion wasn't so different from what most people saw in electrical phenomena for the next two thousand years. Saint Augustine was left speechless by a simple parlor trick of moving bits of iron around a table by using a lodestone concealed underneath. He often recalled the episode as an example of a miracle, definitive proof of the divine. More than a thousand years later, van Helmont was not shy in using the word "magic" to describe the phenomenon of magnetism.

By the nineteenth century, the belief that electricity constituted some secret ingredient of life had become widespread, and it fit neatly into the theory of vitalism. Vitalists maintained that an unbreakable barrier stood between life and nonlife, that living and inanimate matter were fundamentally different and incompatible. To the vitalists, spontaneous generation was simply not plausible without the infusion of an *élan vital*.

Vitalism was an old theory. It could be traced all the way back to Thales's time, and it had a long history in Western medicine, where it found a place in the works of such seminal figures as Hippocrates and Galen, who believed, long before the discovery of air, or even of gases, that the lungs worked by drawing on a mysterious supernatural energy that Thales called *pneuma*. But like the theory of preformation in the time of Needham and Voltaire, vitalism had been reinvigorated—galvanized, one might say—by those who feared the growing threat of materialism.

The nineteenth century was the age of industrialism, the age of the machine. Cities were filling with factories and their endless plumes of smoke. Railways sprawled through once pristine countryside. Everything from architecture to social traditions seemed to be under siege by the rapid progress brought by the industrial revolution. In the sciences, the vision of a universe arranged like a mathematically precise clock, expounded by Newton and Descartes, was growing ever more prominent. The line between the living and the nonliving was blurring. This trend was unsettling for many. Vitalism was an attempt to stave off what its adherents saw as a cold and dehumanizing vision of the universe.

Central to the theory was the idea that living things were different from the nonliving because they had a soul. But what exactly was a soul? The concept needed a scientific veneer fit for an age of scientific skepticism. Believers in vitalism began describing a kind of cosmic fluid. Some called it "ether," a "vital force," or an "*élan vital.*" Still others called it an "imponderable fluid," which one vitalist writer described as an "electric, magnetic-mineral, or organic fluid." The name "imponderable fluid" was a little ironic. It might indeed be invisible, but it held characteristics that could be observed, or pondered. That was the point of vitalism. The seemingly magical pull of a magnet, the invisible force of an electric current—these were observable phenomena that seemed to defy materialistic explanation. This is what set vitalism's "soul" apart from the soul that most people might imagine in the twentieth century. Vitalists believed the soul could be observed, and perhaps measured, if only with great difficulty.*

---

* In 1928, the German chemist Friedrich Wöhler had stumbled upon a chemical reaction that enabled him to convert ammonium cyanate into urea. This first synthesis of an organic compound is often

So eager were people—even natural philosophers—to embrace vitalism that Alexandre Bertrand, the science reporter for the Paris-based newspaper *Le Globe*, wrote of a "revolution in the high regions of physics. . . . The universe appears to us now as if entirely plunged into an infinite ocean of imponderable matter." In hindsight, it would be easy to dismiss the movement as quackery. But in the first half of the nineteenth century, vitalism was so pervasive that it inspired the separation of the field of chemistry into two separate branches, organic and inorganic. Most of the leading figures in the life sciences would have described themselves as vitalists—even Louis Pasteur. They believed in the unassailable line between the living and nonliving. Not all of them believed in "imponderable fluid," but many still saw clues to the nature of life in electricity and electromagnetism.

ANDREW CROSSE'S first connections to the world of electrical science came through his family. His father, Richard Crosse, was a good friend of two men who understood the science of electricity as well as any others in the late eighteenth century did: Benjamin Franklin and Joseph Priestley. The friendships stemmed from the radical politics shared by all three. Richard Crosse was a well-known supporter of the French Revolution and had joined the crowds on the day of the storming of the Bastille, even hoisting the French *Tricolore* over the battlements. His efforts had ruined his reputation in England, where he was seen as an eccentric troublemaker or, worse, a Jacobin revolutionary. Upon his return from France, angry mobs had tried to attack his carriage. Yet Richard Crosse's radicalism also earned him admirers, among whose ranks were Franklin and Priestley. Both men were guests at Fyne Court, and both were scientific visionaries who left important marks on the emerging field of electricity.

At the time, scientists explained phenomena such as magnetism as the product of two distinct and different electrical fluids having two different powers: attraction for one, repulsion for the other. Franklin still believed that electricity was a fluid, but a single fluid, with both positive and neg-

---

mistakenly held up as the end of vitalism, but in reality very little significance was placed on the discovery at the time. Even Wöhler himself remained a vitalist.

ative charges that explained its strange properties. He didn't understand that electricity's apparent movement was simply the flow of electrons between atoms. Still, his view was a huge step toward understanding how electricity worked and what it actually was. While Franklin was serving as the American ambassador to France, his reputation in the field earned him an appointment to a royal commission to investigate claims by a vitalist, the German hypnotist Franz Mesmer, that he could heal people with invisible electrical fluid. Mesmer's cure was to have his subject swallow pieces of iron and then to attach magnets to the rest of the subject's body. The phrase "animal magnetism" comes from Mesmer's belief that electromagnetism constituted a supernatural life-giving force.

Priestley was almost as remarkable a figure as Franklin. Raised in a strict Calvinist household, he became a dissenting clergyman at an early age, turning to Unitarianism and denying the divinity of Christ, and fleeing to Pennsylvania when anti-French rioters burned down his house in England. He is probably best remembered for his work in chemistry and his discovery of oxygen, which he called "dephlogisticated air." Priestley also was the first person to describe electrical force mathematically, in a formula he subsequently included in his seven-hundred-page book *The History and Present State of Electricity*, which became the standard text in the field for over a century.

IN HIS LATER YEARS, Andrew Crosse never mentioned his father's famous friends when retracing his own fascination with electricity. This omission was not surprising. Andrew Crosse was a progressive, but he was merely a proponent of reform, not a radical or revolutionary like his father had been. Andrew Crosse's outlet was the reformist Whig Party, and he even served a term as a member of Parliament representing Somerset County. With Richard Crosse's political views growing ever more marginalized as the years went by, it is not surprising that Andrew Crosse chose to steer clear of reference to Franklin and Priestley to avoid conjuring the specter of his father's radicalism.

Andrew Crosse received his first electric machines when he was sixteen, the year his father died. Crosse was by that time devouring entire volumes of the *Philosophical Transactions* as soon as he could acquire them. He read whatever he could find on the subject of electricity. The bookseller at the shop Crosse frequented turned out to be an experimentalist himself and took an interest in the boy. He gave Crosse a crude generator that could produce energy from simple friction and a "battery table" that held thirty Leyden jars. Little more than a glass jar filled with water and containing a piece of metal foil to conduct electricity, the Leyden jar, named after the Dutch city where it was invented, was one of the earliest versions of an electric capacitor.* In time, Crosse would find a more efficient way to harness electricity from the atmosphere in the condensers and lightning rods he arranged in the trees outside his home. But the Leyden jar remained a staple in his laboratory. Eventually, his basement held three thousand of them.

That same year, workers in a nearby limestone quarry stumbled upon the entrance to a fantastic cave filled with aragonite crystals, which would come to be called Holwell Cavern. As a teenager, Crosse would spend hours sitting alone and watching the crystals glow in the dim flicker of candlelight, as if emanating some strange energy. He became convinced that their beauty was related to this mysterious force he was learning about, electricity. The crystals appeared to reach for each other, in mirror image, as if drawn together by an invisible force. Crosse guessed that the force was magnetic attraction.

Two years after Holwell Cavern's discovery, Crosse left for Brasenose College, Oxford. Crosse was a loner at heart and found his university years trying. Oxford, he wrote his mother, was "a perfect hell on earth." Many years later, he would say that his years at Oxford had taught him that "ridicule is a terrible trial to the young." He found solace in studying the Greek classics. Crosse always fancied himself a poet. He tended to wax on the beauty of nature, and Holwell Cavern became a favorite subject. In later years, his poetry frequently focused on his melancholia or the religious zealots who would turn on him at the end of his life.

---

* Benjamin Franklin was the first to prove that the electricity was stored in the glass casing, not the water, as people had believed. He also used a Leyden jar in his famous kite experiment, from which we get the phrase "to catch lightning in a bottle."

When Crosse was twenty-one, his mother passed away, and he returned to Fyne Court. He found himself the master of a large estate with tenant farmers to manage. When it came to business, Crosse was a disaster. At some point, he was swindled out of much of his wealth, although he was rich enough that he never quite fell into a desperate state of affairs. As the years passed, he turned ever more eagerly to his electrical experiments, encouraged by the man who had by then become his closest friend, the electrical scientist George Singer.

When he was twenty-seven, Crosse devised his first electrical experiment, which he carried out at Holwell Cavern. He even began experimenting on local farmers who came to him with various ailments. It was said he could cure arthritis and even hangovers. Soon he was investigating all manner of possible uses for electric currents. Crosse's electrical investigations into the formation of crystals prompted a pilgrimage to Fyne Court by one of Britain's most celebrated scientists, Humphry Davy, president of the Royal Society. Davy had become a national hero in England after his invention of a lamp that could be used safely in methane-heavy coal mines. He was also a brilliant chemist who had used electricity to great effect in his chemical explorations, and his interest in Crosse's work stemmed from his own experiments. Using an electric current from a voltaic pile, Davy had discovered the process of electrolysis, enabling him to separate substances into their component chemical elements.

By 1836, interest in electricity was booming in England's scientific community, and Crosse was gaining in prestige, as well as in confidence. In the fall of that year, he traveled to Bristol to deliver a rare public address before the newly formed British Association for the Advancement of Science. His theories on the electrical formation of minerals were hailed as visionary and brought him enough notoriety that anyone involved in science knew who he was, as did many with no particular ties to science at all. By the time Crosse undertook his most famous experiment later that year, his scientific career was at its pinnacle.

Later, after events turned out very badly for Andrew Crosse, he would often protest that he was little more than the victim of unscrupulous newspapermen. The truth is somewhere in between. While he did at first keep the discovery of his insects to himself, eventually he confided the story to the editor of a new local newspaper, the *Somerset Gazette*. He could hardly have been surprised that an enthusiastic story was published shortly thereafter.

On December 31, 1837, the first story about Crosse's fantastic insects appeared in the *Gazette*, under the title "An Extraordinary Experiment." Eventually, the story reached London and the paper of record of the day, the *Times*. From there, news of the real "Dr. Frankenstein" went into syndication and spread like wildfire to all of the British Empire and beyond. Soon, newspapers were reporting—falsely, it would turn out—that Michael Faraday, the most famous electrical scientist of his time, had confirmed Crosse's results in his own laboratory. The press gave Crosse's insects a proper Latin species name. They called them *Acarus crossii*.

Crosse tried his best to remain aloof from the celebrity, and he distanced himself from any larger implications of what had happened in his organ-room lab. His subsequent experimental work continued to focus on finding a way to make crystals using electricity. His few follow-up attempts to understand how the insects appeared in his apparatus were fruitless.

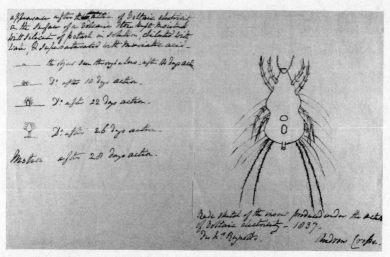

Crosse's diagram of what would be called *Acarus crossii*.

One scientist—William Weeks, a popular lecturer with a large public following—claimed to have replicated Crosse's experiment and found the same insects. But others who tried found nothing. The deeply religious Faraday, though sympathetic to Crosse, denied the reports that he himself had replicated the experiment. Faraday's protestations were virtually ignored in the press, which didn't want to diminish the sensational story, but in the scientific community, they were taken as just more evidence that the experiment was faulty. Crosse's story might have ended there.

IN 1844, CROSSE ONCE AGAIN found himself in the public spotlight when a new anonymously authored work appeared on the shelves of England's booksellers. It was called *Vestiges of the Natural History of Creation*, and the British public had never seen anything like it. The novelist and future prime minister of England, Benjamin Disraeli, excitedly wrote his sister to tell her of the book that was "convulsing the world." The first edition sold out within days.

Vestiges was a naturalistic history of the universe, beginning with the creation of the stars and the heavens and running right up to the present time, fueled by what the unnamed writer called "the universal gestation of nature." The history of life on Earth was traced to an original act of spontaneous generation, with Andrew Crosse's *Acarus crossii* held up as evidence. From this natural act of creation, Lamarckian transmutation became the engine for the creation of new species. In the original handwritten manuscript, next to the section on the origin of life, the author scrawled in the margin that the "great plot comes out here."

In Victorian England, the book created a scandal, but the kind of scandal that made its publishers very rich. People rushed to get their hands on a copy. Prince Edward said he read it aloud to Queen Victoria every afternoon at tea. The fact that it was published anonymously only added to its appeal. Newspapers speculated endlessly about the identity of the mysterious author. Some said it was Andrew Crosse. Some said it was Erasmus Darwin's grandson, Charles. Still others suggested the author had to be a woman, many because they thought only a woman could pen such a shoddy, ill-informed work. Maybe it was the radical political economist Harriet Martineau, they guessed, or Ada Lovelace, Lord Byron's daughter, writer of the first machine algorithm, which she had produced for mathematician Charles Babbage's never completed computational machine.*

Not until after his death some thirty years later was the journalist and publisher Robert Chambers finally revealed to be the book's true author. Chambers went to great lengths to conceal his identity, even having his wife copy the original manuscript by hand, for fear his handwriting would be recognized by editors. He burned his notes and kept the manuscript in a locked drawer. Chambers was terrified of the inevitable religious backlash he would face if his identity was revealed. He and his brother owned a publishing house that earned most of its income from producing religious textbooks in their native Scotland. Discovery would have meant financial ruin.

Though Chambers was a religious skeptic, he went to great lengths in his book to insist that the process he described was, at its root, the work of a divine hand. Even the title was a concession. The content of the book still proved too much of a challenge to biblical accounts of creation for the religiously inclined. Even though it sold astoundingly well, after the floodgates of religious indignation opened, *Vestiges* admirers largely kept silent, while its critics were relentless. Perhaps the most scathing review appeared in the enormously influential *Edinburgh Review.* At eighty-five pages, it was the longest article the journal had ever published. The reviewer was the Reverend Adam Sedgwick, the esteemed geologist and vice-dean of Cambridge's Trinity College, who fulminated about the absence of Adam and Eve and the Garden of Eden and ridiculed the notion that human beings came from apes. Sedgwick was among those who were sure the book had been written by a woman, and he devoted much of his sprawling article to proving his case.

Sedgwick didn't spare Andrew Crosse. He sent Crosse a personal letter

---

* Babbage had first announced plans to build the machine in 1837, the same year Crosse's story appeared in the British press. It was a huge machine, meant to perform basic arithmetic. Had it not proved too expensive to complete in Babbage's lifetime, it would have become the first computer. Lovelace constructed her algorithms on punch cards not unlike those that would be used for the first computers of the twentieth century.

admonishing him never to "meddle again with animal creations; and without delay to take a crow-bar and break to atoms" his "obstetrico-galvanic apparatus." Sedgwick wasn't alone. Following the publication of Chambers's book, Crosse was widely vilified, and he became a Dr. Frankenstein in the most pejorative sense. His experiment had usurped the very role of the creator. A deluge of hate mail began pouring in: he was a "heretic," a "blasphemer," a practitioner of "dark arts." Many newspapers took the same tone. In print, he was branded "a reviler of our holy religion" and a "disturber of the peace of families." Local farmers stopped speaking to him. He was blamed for an infestation of locusts near Fyne Court. A reverend with a reputation as something of a fanatic was called in to conduct a public exorcism of the hills surrounding Crosse's estate.

The religious-tinged criticism hurt Crosse, but it was his fall from grace in the scientific community that really crushed him. Before *Vestiges of the Natural History of Creation* appeared, most of England's important scientific bodies simply avoided addressing the nature of Crosse's insects. The subject of life's origin was so sensitive that they would touch upon it only reluctantly. But the elevation of his experiment in this now infamous book meant he could no longer be ignored. He was subjected to an unprecedented level of scrutiny. Every aspect of his experiment was carefully parsed and found wanting. *Acarus crossii*, many of his peers mused, was nothing more than the common dust mite. He became a laughingstock in the scientific community. The nervous attacks of his youth returned with a vigor. He immersed himself in poetry and tried his hand at fiction, but he had never been much of a writer.

Many years later, Crosse was asked to write a comment for a book about the great historical events of the first half of the nineteenth century. He bitterly protested that he was an unwitting victim and that he had never claimed credit for any act of "creation." He had simply been swept up by the conclusions drawn by others, caught in the tide of an increasingly acrimonious debate between those who saw the world and all its workings as explainable by science and those who saw the world through the lens of biblical creation.

DECADES AFTER the general public had forgotten Andrew Crosse, his name was still invoked among scientists. As the world of science grew more professionalized and science became a career rather than a pastime, Crosse came to humorously symbolize the "gentleman scientist" of a bygone era. The mere mention of his name could be invoked as a synonym for unprofessional quackery.*

But the public saga of Andrew Crosse left behind another legacy, one shared by the publications of *Frankenstein* and *Vestiges*. Each had whetted the appetite of a lay public that was hungry for popular science and ready to entertain ideas that might once have seemed too heretical even to imagine. They were the precursors of a scientific literary phenomenon that would culminate in the publication of one of the most famous and influential science books ever written.

The man who was going to write it had returned to England late in 1836, only a couple of months before Andrew Crosse began his experiment. He was a young naturalist who had just completed a long sea voyage that had fatefully taken him to the Galápagos Islands off the coast of South America.

---

* The historian Trevor Pinch has drawn an analogy between Andrew Crosse's *Acarus crossii* and the announcement of the discovery of cold fusion in 1989. Both events were propelled into the limelight by major newspapers of the day, which drew conclusions beyond the experimental evidence. In Crosse's case, it was the *Times*; in the case of cold fusion, the *Financial Times*. The central figures in both stories went through periods of unwarranted celebrity followed by a descent into equally unwarranted ridicule and infamy. Even the apparatuses used in the two experiments were uncannily similar: electrical conductors that were run through solutions containing potassium salts.

BREATHED BY THE CREATOR
INTO A FEW FORMS OR ONE

It is not enough to discover and prove a useful truth previously unknown, but that it is necessary also to be able to propagate it and get it recognized.

—JEAN-BAPTISTE LAMARCK, *Zoological Philosophy*, 1809

IN OCTOBER OF 1835, a British Royal Navy sailing ship made its way out of an old pirate cove on an island at the eastern edge of the Pacific Ocean, off the coast of Ecuador. The HMS *Beagle* was a ten-gun sloop outfitted with an extra mast for an ambitious voyage of exploration and discovery. For the previous four years, the *Beagle* had sailed southwest from Britain, through the Azores and Cape Verde, hugging the coast of most of South America, until finally reaching the Galápagos Islands.

The ship's captain, the young but well-seasoned Robert FitzRoy, had decided to search elsewhere for the freshwater needed for the next leg of the ship's journey, the long trek west across the Pacific to Tahiti. But he left four men behind, to be picked up on the ship's return ten days later. The party included the *Beagle*'s surgeon Benjamin Bynoe, two servants, and the ship's twenty-six-year-old gentleman-naturalist Charles Darwin.

Darwin took any opportunity to escape the *Beagle* and explore on his own, frequently with Bynoe in tow. As the ship naturalist, Darwin was responsible for observing and collecting samples of the flora and fauna they encountered. But sometimes he just needed to get away. Darwin's relationship with the captain could be cantankerous. Often they argued about politics. FitzRoy was an impassioned Tory; Darwin, a committed Whig. Sometimes they bickered about slavery. Darwin's grandfathers were two of England's most prominent abolitionists—Erasmus Darwin and Josiah Wedgwood—and he adamantly opposed the institution. Usually, though, Darwin and FitzRoy simply got on each other's nerves. Being confined on a small ship for years on end has that effect on people. Yet, as FitzRoy well knew, loneliness could be far worse. The previous captain of the *Beagle*, Pringle Stokes, had taken his own life when FitzRoy was under his command. FitzRoy chose Darwin as much for his suitability as a companion as for his strengths as a scientist. Many years later, the captain would come to regret that decision, after he turned to a form of religious fundamentalism that Darwin would do so much to undermine.

James Island was one of the bigger islands at the center of the Galápagos archipelago. It had been named by the seventeenth-century buccaneer Ambrose Cowley, whose maps FitzRoy still used. Until the ship returned in a week and a half, Darwin's small party would be left to fend for itself.

The ship's crew had first set foot on the Galápagos three weeks earlier, on Chatham Island. FitzRoy described the volcanic beach where they had landed as "black, dismal-looking." Darwin imagined that hell would look much the same. The heat was oppressive, and he had measured the sand on the beach at 137°F. The more abundant black sand, volcanic and hard to the touch, was even hotter. In his journal, Darwin wrote that it was difficult to walk on even "in thick boots."

Volcanoes abound in the Galápagos, and the signs of volcanic activity intrigued Darwin. He was well trained as a geologist. At Christ College, Cambridge, he had taken courses taught by two of the most esteemed geologists of their day, the Reverend Adam Sedgwick—who so attacked Andrew Crosse and *Vestiges*—and John Henslow. The two were his closest mentors. Before the *Beagle* set out, Henslow had recommended a book to occupy Darwin's long days at sea: *Principles of Geology*. The book was published in three volumes, and Darwin had received the first volume from

FitzRoy as a gift. Though Henslow, a clergyman like Sedgwick, had warned him not to take the book too seriously, Darwin couldn't help himself. He was captivated.

The book's author was a brilliant geologist named Charles Lyell. For Lyell, geology meant something entirely different from what it meant to geologists like Sedgwick. Sedgwick was well versed in the clamors and catastrophes of the natural world, earthquakes and floods and the like. But these were, for him, isolated events, and geology was a science based on fixed positions on maps, immutable through all of time. In contrast, Lyell saw geology as a science of change, of natural processes that were constantly reshaping the world, forming entirely new mountains and seas and rivers. Treating these processes as clues, Lyell arrived at his most revolutionary contention of all, calculating that the world was at the least three hundred million years old.

Isolated at sea and immune from other distractions, Darwin immersed himself in Lyell's book. Early in their voyage, on the Island of St. Jago (modern-day Santiago) in Cape Verde, he had come across a layer of shell and coral 30 feet above sea level that contained the petrified remains of mollusks. Darwin saw these as evidence of Lyell's contention that landmasses were rising. Everywhere the *Beagle* went, Darwin seemed to stumble upon the geological processes that Lyell described. With Lyell's observations on his mind, he began to see the world anew. He grew ever more inquisitive, seeking out his own answers, even to questions that Lyell himself supposed he had answered. In the solitude of his cabin after his explorations on St. Jago, Darwin wondered whether, instead of the reefs rising, it was the ocean that was *sinking*.

JAMES WAS ONE OF the biggest volcanic islands in the Galápagos. Because he had found his St. Jago fossils amid volcanic sediment, any sign of volcanic activity in the Galápagos intrigued Darwin. In the three weeks before he arrived on James, he had seen volcanic cones that rose 60 feet in the air like "the Iron furnaces near Wolverhampton." He guessed that the islands might hold two thousand cones like these, and the beach on which the

Beagle had left Darwin's small group was flanked by two such craters, both unusually large. Darwin guessed that their past eruptions had formed the cove itself—that the Earth was living, moving, its convulsions shaping the very island upon which he stood.

After two days of exploring the volcanic beaches of James, Darwin realized he wasn't going to find fossil deposits like those he had discovered at St. Jago. So he set about collecting specimens. Other than the occasional stunted tree that seemed to be clinging just barely to life, there were few signs of indigenous species around the craters. Even insects were scarce. Only tortoises were abundant, great lumbering creatures that could grow as long as 3 feet. These held little interest for Darwin, who assumed incorrectly that the species was not native to the islands, but had been brought by colonizers. They were, however, a great source of food. Until the *Beagle*'s return ten days later, Darwin and his companions survived on almost nothing besides turtle meat fried in turtle fat.

Soon the party began venturing deeper into the interior. One hike took Darwin and his companions about 2,000 feet above sea level and 6 miles from the shore. They encountered a group of Spanish whalers who took them to see a salt lake at the bottom of a crater, the water barely concealing "beautifully crystallized, white salt" and surrounded by "a border of bright green succulent plants." Amid the bushes surrounding the lake, they encountered a local historical curiosity, the skull of a whaling captain who had been murdered by his crew. Two days later, Halley's Comet streaked across the sky. Darwin noted the event in his notebook with but a single word, "comet." It was the Earth, not the heavens, that captivated Charles Darwin.

Darwin had promised his former professors, Henslow and Sedgwick, that he would return with all the plant and animal specimens he could. Both were exceedingly curious about what Darwin would find in the Galápagos. By the time he arrived in the islands, he needed little encouragement. The wildlife of South America had entranced him. Off the coast of Patagonia, the *Beagle* had been surrounded by a vast migration of butterflies so thick that some sailors shouted it was "snowing butterflies." At sea, Darwin had seen phosphorescent jellyfish that glowed in the dark of night. Other nights, they passed over patches of phosphorescence, beautiful and luminous, emanating from deep beneath the surface of the ocean.

In Patagonia, Darwin had purchased fossilized bones of strange and exotic creatures. A fossilized skull looked like it could be the head of giant rat, bigger than anything Darwin had ever imagined. He also came to possess the bones of an animal that seemed to be a large camel, which left him wondering about the natural changes the continent must have undergone to lead to the extinctions of such species. He mused about Buffon's notion from *Natural History*, that the wildlife of the Americas was weak and had no vigor. In his notebook, Darwin wrote that if Buffon could have seen what he saw, surely he would have thought differently.

Away from the volcanoes on St. Jago, Darwin discovered that the island was teeming with life. He gathered almost any plant he could lay his hands on but collected only the most interesting animals, since he had to perform his own taxidermy and had to share his cabin with the carcasses until they could be preserved. Birds were common on the islands and made up the bulk of his collection. Because they were so unaccustomed to humans, hunting them was almost too easy. Darwin once managed to get so close to a hawk that he was able to prod it with the point of his gun.

Nearly everywhere he went in the Galápagos, Darwin found one particular bird species thriving.* In his journal, he called them *thenka*, the Spanish name for birds he had encountered on the mainland in South America, which he had assumed to be mockingbirds. By the time he arrived on James, these birds were beginning to arouse a particular fascination in him. Two weeks earlier, the *Beagle* had stopped at the prison colony on Charles Island. The prisoners had claimed that each island was home to a different kind of tortoise that could be identified by its shell. The vice-governor said he could tell which island he was on by the tortoises alone. Darwin put little stock in the claims. James afforded plenty

---

* Although they would come to be known as "finches," the birds Darwin encountered on the Galápagos were not related to true finches in the taxonomic sense. The confusion is largely the result of the popular 1947 book *Darwin's Finches*, written by David Lack.

of opportunity to study the tortoises up close, and they seemed identical to those he had already seen. But when Darwin began to notice little differences in the mockingbirds, he was reminded of his conversations on Charles. Those of each island seemed to have a beak distinctive from those of the others. Some beaks were much larger or smaller than others. Even the beaks' shapes varied, from narrow and pointed to wide and downward curved. By the time he arrived on James, he had begun recording the exact island where each of his birds originated.

Darwin had by then finished reading the next volume of *Principles of Geology*, which Henslow had sent to him in Uruguay. Lyell was a lawyer by trade, and in his second volume he turned his formidable skills at argument and logic to tearing down the transmutational assertions of Jean-Baptiste Lamarck. Lyell saw in the geology of the Earth a great transformational force, yet he saw no such force when it came to the plants and animals that inhabited it. Lamarck looked at the fossil record as proof that species had changed over time. Lyell contended that what Lamarck was really seeing was evidence of extinctions, followed by acts of creation, where God brought new, more able species into being. These new creations explained what seemed to be rather abrupt changes in the fossil record. There were no natural links between species dying out and new species appearing. The appearance of each new species was a miracle.

In the Galápagos, Darwin began to part ways, ever so cautiously, with Lyell's opinions about transmutation. In the mockingbirds of the islands, he began to see evidence of a gradual transformation of species, shaped by the demands of the changing environment that Lyell assumed. But he wasn't yet quite sure. The birds had differences, but in Darwin's view they were still merely variants of the same species. He noted his confusion in his ornithological journal. If these birds indeed turned out to be not simply variants, but distinct species, "such facts would undermine the stability of Species." Their close proximity on each of the islands and their apparent strong similarity could not be mere chance, but evidence that the birds shared a common ancestor, from which they had evolved. Darwin was beginning to see Lyell's processes of change at work on not just Earth's geology, but life itself. In the Galápagos, he would write, "We seem to be brought somewhat near to that great fact—that mystery of mysteries—the first appearance of new beings on earth." Nineteen years would pass before he explained what he meant by that sentence.

From the Galápagos, the *Beagle* spent the final year of its journey sailing across the Pacific and around the horn of Africa, arriving in England in October of 1836, at the port of Falmouth. A steady stream of accounts of the expedition appeared in newspapers, and Darwin returned to find that his reputation had bloomed while he was away. For five years, he had maintained a correspondence with Henslow, and his geological observations so impressed his former professor that Henslow had gathered them into a pamphlet that he shared with other naturalists. Lyell was one of those who received a copy. He couldn't wait to meet the young man who described himself as Lyell's "disciple."

Within a month of Darwin's return, the two dined at Lyell's London home. They got on famously from the very beginning, forming a friendship that would last their lifetimes. Lyell listened in rapt fascination as Darwin regaled him with stories of earthquakes in foreign lands. Lyell offered to shepherd the young man in his new career. He also advised Darwin to stay in London, where Darwin would be close to the array of specialists whose help he would need in deciphering all that he had seen and learned. Lyell invited one of the specialists he had in mind to dinner that evening, a young anatomist named Richard Owen, who would four years later coin the term "Dinosauria," or "terrible reptiles." Before Darwin left that night, Lyell, the head of the Geological Society of England, offered him one last piece of advice that seemed premature: waste no time heading a scientific organization. He was already certain that Darwin would go far.

Lyell and Darwin became nearly inseparable. For a spell, it seemed as if they met every day. Lyell helped guide his new protégé as best he could. Memberships in the most esteemed scientific organizations started piling up, including the Royal Society, the Royal Geographical Society, and, of course, Lyell's Geological Society. Darwin also received a royal grant to help him write an account of his adventure and his observations that would be published as part of a multivolume memoir that FitzRoy had planned.

Darwin began putting his fieldwork data in order. To catalog the fossils he had found in Patagonia, he enlisted Richard Owen's help. Boxes of Darwin's fossils began arriving at the Royal College of Surgeons, which Owen headed. Darwin could still only guess as to what exactly the odd collection of interesting-looking bones would eventually yield. Owens soon informed Darwin that his collection contained pieces of a giant llama and the head of gigantic rodent that would have been about the size of a hippopotamus. Lyell gloriously paraded his protégé's finds before the Geological Society in an exhibit he called "Darwin's Menagerie." For Lyell, these fossilized bones were evidence of a fascinating "law of succession," in which God arranged his creations on the planet Earth in a kind of geographic order. Though similar in shape and structure, each was still unique, and unrelated by ancestry.

Darwin, however, was already beginning to suspect that all of his animal specimens were merely branches of a single family tree, each related by blood and genealogical history. These were still just speculations, but his instincts told him the giant llama had to be a cousin at least of the llamas that still lived on the South American continent. There had to be some connection, and an explanation that could be found in the decipherable laws of nature. It was not enough to simply see these variations as the work of divine creation. Owen, for his part, was a committed vitalist who believed that all living things contained an inherent "organizing energy" that governed bodily processes like growth and decay. Conservative and religious, he was, like Lyell, an ardent opponent of evolutionary concepts. Both would eventually change their minds. Owen would swing so far in the other direction as to one day attack Darwin for not embracing the implications that his evolutionary theory held for the origin of life.

In MAY OF 1838, FitzRoy's four-volume account of the expedition was published under the title *Narrative of the Surveying Voyages of His Majesty's*

Ships Adventure and Beagle. The first two volumes consisted of FitzRoy's memoirs of the *Beagle*'s expeditions, including the ship's first voyage under the command of Stokes. The last volume was a lengthy appendix. Darwin's account made up the third volume and quickly began to outstrip the others in popularity. It was soon published on its own as the *Journal of Researches into the Geology and Natural History of the Various Countries Visited by H.M.S. Beagle.* Later, it became, simply, *The Voyage of the Beagle.*

The description of the visit to the Galápagos filled only a small portion of Darwin's account, but it was the most historically significant. Of particular interest was a sentence about some birds he by then believed to be finches. "One might really fancy that from an original paucity of birds in this archipelago," he wrote, "one species had been taken and modified for different ends." It was a tantalizing—if cautious—glimpse of his increasing doubts about the notion of fixed species.

The animal specimens Darwin brought back took longer to decipher than his fossilized bones. In all, he brought back 80 stuffed mammals and 450 birds, which he entrusted to the Zoological Society of London. Despite having just added a new museum in the city's posh West End, the Zoological Society took them only reluctantly, and processing was slow. The society was turning into an organizational nightmare. It had put out a call for submissions for exhibitions but was unable to cope with a torrent of animal specimens that seemed to arrive daily from hunters and naturalists around the world, all of which needed to be tagged, described, and stored. Eventually, though, Darwin's specimens found their way into some of the surest hands the organization had, those of the taxidermist John Gould. A self-educated former gardener, Gould had pulled himself up through his skill at taxidermy, becoming the Zoological Society's first curator and preserver. He was also a fine painter who had authored and illustrated several popular books on birds.

The little birds Darwin had retrieved from the Galápagos were of particular interest to Gould, just as they were to Darwin. Gould concluded that these birds were actually thirteen distinct species of finches and three of mockingbirds, none of which existed outside their respective islands in the Galápagos. The differences in their beaks reflected the food sources unique to each island: some were perfect for consuming various types of cactus seeds; others, for eating insects. Darwin finally had the answer he'd been waiting for, and he was confronted by evidence of evolution even more striking than that given by his fossils. Darwin was sure that each bird was related, that they all shared a common ancestor. At some point, a species of finch had made its way to the Galápagos and transformed itself into thirteen separate species. Only their beaks had adapted to the environment. But how? He stumbled upon a clue in the most unlikely place, an essay on political economy that had been suggested by a friend of his older brother Erasmus.

FREETHINKING ABOLITIONIST, portly poet, outgoing and generous, Erasmus Darwin was in many ways the mirror image of the grandfather for whom he had been named. Though he and his brother Charles differed little when it came to subjects like politics and religion, Erasmus was always the more daring. Like their grandfather, he was willing to strain the boundaries of respectability. Also like their grandfather—and their father—he was a physician, though he suffered from the same chronic bouts of illness that would one day haunt his brother. Their father, fearing that his son's medical career was too severe a strain on Erasmus's "body and mind," advised his son to take early retirement. By the time Charles returned from the *Beagle* expedition, Erasmus had followed their father's suggestion and given up practicing medicine. He had not yet turned thirty.

But Erasmus remained active in society and in Whig political circles. Money was no impediment. Both he and Charles had at their disposal the large family fortune that stemmed from the industrial successes of their maternal grandfather, Josiah Wedgwood. Though Erasmus would eventually die a bachelor—and with an opium addiction—he was linked romantically to a number of freethinking women, including the radical political economist Harriet Martineau, who was mistakenly implicated as the mysterious author of *Vestiges of the Natural History of Creation*. The two began a long affair, and a lifelong friendship. The prospect of marriage to the outspoken, modern Martineau worried Erasmus's father, who asked Charles to keep an eye on his brother. Over the coming years, Charles became, if not her friend, a close acquaintance.

Harriet Martineau was one of the most prominent intellectual disciples of the Reverend Thomas Malthus. A former economist for the British East India Company, Malthus had argued that increasing population growth would inevitably lead to increasing poverty, as the labor surplus would lead to a massive reduction in wages. The weak and the poor would be weeded out in the struggle for survival that was the natural state of human society. Starvation, disease, war, and even infanticide were, in Malthus's vision, natural checks on this delicate economic balance. His ideas earned him a place as the most influential economist of his time and spawned a reform movement that eventually led to a strengthening of the English "Poor Laws," which established a system of workhouses for the most indigent.

Malthus's seminal book was *An Essay on the Principle of Population*. After a number of conversations with Martineau about Malthus, Darwin got around to reading the book in late 1838. Almost immediately, he had an epiphany. The struggle that Malthus identified in human society was the same as the struggle for survival that Darwin observed in nature. Darwin would later write in his autobiography, it "at once struck me that under these circumstances favourable variations would tend to be preserved and unfavourable ones to be destroyed. The result of this would be a new species. Here then I had at last got hold of a theory by which to work." In his private notebooks, Darwin began reexamining the earlier theories of transmutation that had enraptured his grandfather. By applying Malthusian principles to the natural world, he finally grasped the basic mechanism of transmutation that had eluded his predecessors in the world of evolutionary theory. The variations in living creatures were the product of natural selection.

In CHARLES DARWIN'S TIME, most people—including most scientists—saw life as something that had changed little since the era of biblical creation. Joseph Priestley, himself no orthodox follower of scripture, summed up the conventional view when, in a book about spontaneous generation, he wrote that the "plants and animals described in the book of Job are the same as they are now, and so are the dogs, asses and the lions of Homer. The world is, no doubt, in a state of improvement, but notwithstanding this, we see no change in the vegetables and animals of other species."

By the early nineteenth century, theories of evolution, then generally known as transmutation or transformationism, had grown among the fringes of accepted science. Darwin was no stranger to the basic concept. His grandfather Erasmus held that organisms were slowly but constantly changing, until they became entirely new species. Yet years of relentless religious criticism had left Erasmus Darwin largely silent on the subject by the time his grandson Charles entered the picture. Most of Charles Darwin's introduction to transmutational theory had come at university in Scotland, at the hands of a freethinking biologist named Robert Grant.

Before Cambridge, Darwin had studied at the University of Edinburgh, where both he and his brother had worked toward degrees at the university's medical school, which then had a reputation as the finest in the country. There, Grant took Darwin under his wing. The University of Edinburgh also had a reputation for radicalism, and Grant—Francophile, self-proclaimed enemy of the church and tradition, and, some said, a homosexual—was one of its most radical professors.

Darwin's first impression of Grant was of a man stiff and aloof, though he soon found the professor to be quite the opposite. Warm, friendly, infectiously enthusiastic, Grant could work himself into a frenzy when discussing his scientific passions, microscopy in particular. Grant took his young charge on frequent strolls along the shore near a home he owned on the North Sea. As they gathered sea slugs and sea mats and dark-green seaweed that they called "dead man's fingers," Grant's passion for nature started to rub off on Darwin.

The outspoken Grant had by then become one of the loudest English proponents of the evolutionary concept. Among trained physicians like Grant or Darwin's grandfather, such notions were slightly more acceptable than in other branches of science. These were men who spent years studying bones and vital organs, and bones and vital organs tell a story. Each species, so different on the surface, often shares the same internal body plan. The bones of a rat's hand are remarkably similar to the bones of a human being's hand; just as a bird's wing is similar to a dolphin's flipper. When the skeletal remains of different species are laid out side by side, their relationships can be seen, each species differing only slightly from those nearest to it in the evolutionary chain. Those same similarities could be found in the vital organs. The heart of a bear, for example, differs only slightly from the heart of a cow. These were clues of long, incremental change, beginning with the smallest and simplest organisms and ending with the largest and most complex.

Grant's intellectual hero was the Frenchman Jean-Baptiste Lamarck. A soldier-turned-botanist, Lamarck had devised the first truly comprehensive theory of evolution. Others had vaguely articulated the concept in one form or another. Even several of the classical Greek natural philosophers had guessed at the core tenets of evolutionary theory. But Lamarck laid out a mechanism to explain the process: the theory of acquired characteristics. He held that living things could pass on to their offspring the traits they developed during their lifetimes. Increasing one's speed by running would lead to faster children; the lifting of heavy objects would lead to stronger ones. This natural process of change could be traced all the way back to the spontaneous generation of the simplest building blocks of life, particles that Lamarck called "monads." Spontaneous generation, Lamarck supposed, was happening constantly, providing new evolutionary lines to replace those that had become extinct. While Darwin would one day envision a tree of life, Lamarck saw each species as the product of its own distinct line. Lines that became extinct were replaced by new lines that began by the constantly occurring process of spontaneous generation. Thus, he believed that the most complex organisms—like humans—were also the oldest; the simplest organisms, like protozoa, the youngest.

Darwin's stay in Edinburgh and his molding at the hands of Grant were cut short prematurely. Stories had started to appear in the press of the growing subversiveness at the university. One lecturer generated a furor by suggesting that consciousness was but a function of the natural processes in the brain, that there was no such thing as a soul. Alarmed by what he was hearing about Edinburgh, Darwin's father began to chart a different life for his youngest son. Charles, he decided, would be a parson. And he would prepare for his new life at the more conservative Cambridge, where the professors had more staid notions about the nature of life than did Robert Grant.

Cambridge was hardly a bastion of conservatism when it came to modern science. But compared to Edinburgh, it must have seemed like one. As part of his core curriculum, Darwin was expected to read and digest the work of Archdeacon William Paley, who argued that living beings were so supremely fitted to their environments that this alone could prove the existence of God. God was not quite an active agent, constantly intervening with miracles, but rather a great cosmic designer. Paley liked to use an analogy similar to the one Voltaire was so fond of. "The watch must have had a maker," he wrote. "There must have existed, at some time and at some place or other, an artificer or artificers who formed it for the purpose which we find it actually to answer, who comprehended its construction and designed its use." Paley's views were held by most respectable scientists of the time, and taught by men like Darwin's Cambridge professors Sedgwick and Henslow. When Darwin read Paley's book *Natural Theology*, he wrote that he found himself "charmed and convinced."

By the time of his graduation, his father's idea of a country parsonage had begun to appeal to Darwin. He envisioned a life spent leisurely tending a Unitarian flock, leaving him plenty of time to pursue naturalism and writing. But while journeying around the world in the *Beagle* and beginning to confront uncomfortable truths about the natural world, the idea of a parsonage had steadily lost its appeal. His sisters had guessed as much from his letters home. In a letter that greeted him at his first stop in Falmouth, his sister Fanny wrote, "I fear there are but small hopes of your still going into the Church."

Darwin did indeed return from the voyage of the *Beagle* a changed man. Yet he still sought the approbation of the respectable scientific establishment. He now had little time for the rabble-rousing, irreligious Grant. Back in England, Darwin spurned Grant's overtures in favor of Henslow's and Sedgwick's. Darwin wanted to build a stable life and a stable reputation. In 1839, he married his cousin, Emma Wedgwood, whom he had known since childhood. They eventually took a house in the village of Downe, in Kent. He found that the quiet life of an English village appealed to him. It was like the country parsonage he had once imagined himself heading, but without the flock to tend to.

Religion always remained a delicate subject in Charles and Emma's relationship. Although her Unitarianism wasn't based on literal readings of the Bible, Emma was nonetheless a woman of deep and sincere religious conviction. Charles was already beginning to feel the sway of doubt by the time they decided to marry. Before proposing, he had confessed his growing belief in transmutation. She was, in fact, the first person to whom he confided his increasingly unorthodox views. She wrote to him afterward, saying she was concerned about the doubts raised by his scientific explorations and worried that religion would be a "painful void between us." She accepted his proposal nonetheless. Their marriage was mostly a happy one.

As the years passed and Darwin was increasingly forced to confront the implications of his scientific work, his doubts grew stronger, and stronger still as three of their ten children perished at young ages. By the end of his life, the idea of a divine creator would begin to ring hollow to him. Life did not follow a plan. It did not make sense. He never called himself an atheist, nor did he feel a kinship with the radical atheists who were the quickest to rally to his evolutionary theory. Eventually, though, he did start to call himself agnostic. His beliefs were an obstacle that he and Emma were able to overcome. In public, he would touch on the subject of religion only reluctantly.

Married and settled at Downe, Darwin set about explaining a new theory of evolution, one in which the engine of change was his own concept of natural selection, not the acquired characteristics proposed by Lamarck. He gathered his thoughts into an essay that laid out the basic points of his theory. For years, he showed it to no one. He kept the manuscript in a sealed envelope that he entrusted to his wife. He told her to have it published in the event of his death.

In 1844, Robert Chambers's *Vestiges* was published. The book both excited and alarmed Darwin. To his mind, it was filled with hasty generalizations. Yet it was an articulation of an evolutionary history of species not unlike what he himself was contemplating. And its success proved that the public was thirsty for new ideas about the natural world, even when those ideas flouted convention and religious sensibilities.

Darwin had by then confided his growing belief in evolution to a few close friends, including Lyell and Owen. Both slowly started coming around to his position. Another of his confidants was a young botanist by the name of Joseph Hooker, to whom he showed a draft of his essay, which had grown to 230 pages. Over the coming years, Hooker would become Darwin's closest friend, and a trusted sounding board for his ideas.

Darwin began laying plans for his magnum opus on the subject. He intended to call it *Natural Selection*. The massive sales of Chambers's book showed Darwin just how popular his own book could be. But he was wary of the virulent criticism of *Vestiges*, with some of the harshest attacks having come from his former professor, Sedgwick. Already, Sedgwick seemed to sense the change in Darwin and was becoming distant and aloof. This worried Darwin because Sedgwick was just the type of man he hoped to cultivate and, in time, convert. Darwin wasn't interested in science on the fringes, preaching to radical scientists like Grant. Darwin became even more determined to arm himself with research that was exhaustive and, as much as humanly possible, unassailable.

The work was painstakingly slow. For more than a decade, he gathered material, while working on subjects that would help him build a case. He wrote an intensive study of barnacles and began breeding pigeons in the hopes of finding more clues to and evidence of evolution.

As Darwin forged ahead, there was one aspect of his burgeoning evolutionary theory that Hooker, in particular, had started to fret over: the source of the first living being on Earth, what Hooker called "the vital spark." In the wake of *Vestiges*, the subject of the origin of life was increasingly being raised in the newspapers. In scientific circles, men like Robert Grant continued to fulminate against biblical creationism. For Hooker, it was the ultimate question. Had it been an act of miraculous creation or a natural process like all the other elements of evolution? Lyell was similarly tormented. If Darwin was right, humankind would lose what Lyell called its "high estate."

Darwin shrugged off the need to challenge such entrenched beliefs. There was too little evidence. Any attempt to address the ultimate origin of life would threaten the airtight case he was building for natural selection. He was already tormented by the religious implications posed simply by embracing the transformationist unorthodoxy. He dreaded coming to be seen as a "Devil's Chaplain," the nickname of the evangelist reverend-turned-atheist Robert Taylor, who used to haunt the streets around Cambridge giving fiery speeches. Darwin was wary of anything that would cloud his message or make it harder to accept.

THE STRESS OF WORKING on what he knew would be such a controversial—and important—book took a toll on Darwin. He once compared the admission of his evolutionary beliefs to "confessing a murder." His health deteriorated, probably because of the mental strain. He started to worry about an early death, increasingly experiencing heart palpitations and chest pains that he had first felt on the *Beagle* expedition. He developed a chronic stomach condition he diagnosed as "nervous dyspepsia." Doctors usually advised him against working too hard. He grew ever more retiring. In his autobiography, he wrote, "We went a little into society . . . but my health almost always suffered from the excitement, violent shivering and vomiting attacks being thus brought on."

Darwin continued his work, but at a snail's pace. Then, in June of 1858, he received a jolt in the form of a package delivered from the island of Ternate, in the Dutch East Indies. It contained a handwritten essay by a young naturalist named Alfred Russel Wallace. Shockingly, it looked almost identical to the essay on natural selection that Darwin had written in secret more than ten years before.

Alfred Russel Wallace was a self-made man in almost every respect. He had worked as a carpenter, surveyor, and teacher, educating himself in public libraries, before boldly setting out to follow in the footsteps of his heroes, naturalist explorers like Alexander von Humboldt and Darwin. That Wallace managed to travel the world independently was a feat in itself. Inspired by a popular travel memoir by the New Yorker William Henry Edwards, *A Voyage up the River Amazon*, Wallace first set out for Brazil. He supported himself by sending specimens back to wealthy collectors and museums in England. On his return journey, he narrowly escaped a fire on the ship he was sailing on, surviving for ten days in a lifeboat on the open sea. His next trip took him to Indonesia, where he collected specimens for Darwin, among others.

Wallace was first exposed to evolutionary thinking by *Vestiges*, which had a profound effect on him. Later, in the Amazon, he observed, firsthand, morphological differences in butterflies, much like those that Darwin noticed in the finches of the Galápagos. Also like Darwin, Wallace had read Malthus's influential book on population growth and managed to connect these Malthusian ideas to the struggle for survival that he saw in the wild. Now, Wallace was sending a draft of those ideas to the man he thought could understand his theory and whose approbation he most desired.

Darwin was in shock. He fired off letters to Lyell and Hooker, who had prodded him for years to publish quickly, lest his discovery be scooped. Now that he faced just such a scenario, his mentors came up with a plan to solve his dilemma: they would announce the discoveries of Darwin and Wallace simultaneously. Two weeks later, they presented the findings of both men at a meeting of the Linnean Society of London, where the announcement went almost unnoticed. It took three months for Wallace to hear of the arrangement. He accepted the news gracefully, though he had been left with little choice.

Despite having just lost a third child in infancy, a boy named Charles, Darwin rushed to complete his book. Within thirteen months he finished a compact version of the opus he had originally set out to write. It had a new title, *On the Origin of Species by the Means of Natural Selection*, shortened in later editions to *On the Origin of Species*. Though it sold extremely well, Darwin's book, never managed to outpace sales of *Vestiges*, despite the fact that the latter had already been in print for fifteen years.

From a historical perspective, *On the Origin of Species* was probably the most influential science book ever published. It was written in the same personable, autobiographical style found in Darwin's account of the *Beagle* expedition. His son later noted the book's "simplicity, bordering on naiveté," which may have softened the impact of the new theory. In places, the book was eloquent, but at root it was, in Darwin's own words, "one long argument." It was a lawyer's argument in the style of Lyell's *Principles of Geology*, where evidence was heaped upon evidence. Possible objections were anticipated and shot down before they could take root.

Darwin's description of the history of life finally saw the light of day on the twenty-fourth of November, 1859. According to Darwin's vision, individual species arose through the process of natural selection. Gone was the miraculous divine creator. In its stead was nature "daily and hourly scrutinizing, throughout the world, every variation, even the slightest; rejecting that which is bad, preserving and adding up all that is good; silently and insensibly working . . . at the improvement of each organic being in relation to its organic and inorganic conditions of life."

From the day of its publication, the book generated a firestorm of publicity. Letters started to pour in to Darwin's home. Most of them were critical, but many contained praise. One of those who wrote was the novelist Charles Kingsley, a socialist and a priest in the Church of England, who said the book left him in awe. "If you be right," he wrote, "I must give up much that I have believed." The strength of Darwin's arguments gradually won over the scientific community. The transmutation of Lamarck, so radical and threatening, became the evolution of Darwin, so rational and coolly objective.

DARWIN DID NOT deal with the "vital spark" Hooker had fretted over until near the book's end. Even then, the origin of life was treated almost as an afterthought. "There is grandeur in this view of life," he wrote, "with its several powers, having been originally breathed into a few forms or into one; and that, whilst this planet has gone circling on according to the fixed law of gravity, from so simple a beginning endless forms most beautiful

An 1882 cartoon showing the implications of Darwinism.

and most wonderful have been, and are being evolved." In a second edition released the following year, the phrase "having been originally breathed into a few forms or one" became "having been breathed by the Creator into a few forms or into one."

There is a certain logic to the fact that Darwin's book did not delve too far into the notion of the origin of life. It was a book based on careful observation, crafted to navigate the criticisms that had once been heaped upon *Vestiges*. It was not a speculative book, and Darwin understood that his private beliefs on the subject of life's ultimate origin would be just that. The phrase "by the Creator" was added to mollify Darwin's critics in religious circles. Darwin would come to regret having made the change, and it would be pulled from the third edition.

The way Darwin dealt with the origin of life was disappointing to many in the scientific world, including some in the close circle that was forming around him. They could see in Darwin's theory the same implication that Darwin's religiously minded critics would: life was completely the product of the forces of nature, nothing more. Even in the editions of *Origin* that didn't contain the explicit reference to a "creator," the allusion to life being "breathed into" something was easy to interpret as more than just overly cautious science. It smacked of cowardice.

One of those critics was Richard Owen. If there was a living embodiment of just how far *Origin* had carried evolution into the mainstream of science, Owen was it. Staunchly conservative and religious, Owen had once sneered at the very idea of transmutation. Within a few years of the publication of *Origin*, Owen began sneering at evolutionists like Darwin who he felt had not gone far enough.

In 1863, Owen published a book review in the pages of the *Athenaeum*, one of London's most important literary magazines. The subject was ostensibly a book on microscopic organisms that delved into the issue of spontaneous generation, but Owen used the forum to throw criticism Darwin's way. Owen's name had been left off the masthead, but Darwin immediately recognized his old friend's caustic style and the ideas he knew Owen held.

In his article, Owen criticized Darwin for ignoring the body of evidence that microscopic organisms could arise spontaneously in mud, which would have made *Origin* a more complete explanation for the existence of life. He also took Darwin to task for his remark that life had been mystically "breathed" into the first organism, ridiculing Darwin for having described the first appearance of life in what Owen, relying on a once widely used expression referring to the first five books of the Old Testament, characterized as "Pentateuchal." Owen continued the argument three years later in his book *On the Anatomy of Vertebrates*: "The doctrines of the *generatio spontanea* and of the transmutation of species are intimately connected. Who believes in the one, ought to take the other for granted, both being founded on the faith in the immutability of the laws of nature."*

Darwin was apoplectic. In a letter to the editor published in the *Athenaeum*, he asked, "Is there a fact, or a shadow of a fact supporting the belief that these elements, without the presence of any organic compounds, and acted on only by known forces, could produce a living creature?" But when it came to his choice of words on the appearance of the first life, Darwin became almost apologetic, saying that in "a purely scientific work I ought perhaps not to have used such terms; but they well serve to confess that our ignorance is . . . profound."

Even before Owen's review appeared, Darwin was expressing much the same contrition about his word choice. "I have long regretted that I truckled to public opinion, and used the Pentateuchal term of creation, by which I really meant 'appeared' by some wholly unknown process," he had written to Hooker. Meanwhile, Darwin's views on the subject of the appearance of the first living organisms were themselves evolving.

By the next decade, with his reputation as a preeminent scientist beyond doubt, Darwin began to speculate more about that wholly unknown process, which he increasingly understood to be some form of spontaneous generation of organic life from inorganic building blocks. Again, his thoughts found their voice in a letter to Hooker. One, written in 1871, would become his most remembered on this topic. In it, Darwin wondered about the conditions that might have given birth to the very first living things. It would be, he guessed, "some warm little pond with all sorts of ammonia and phosphoric salts, light, heat electricity &c. present, that a proteine compound was chemically formed, ready to undergo still more complex changes." It was a strikingly modern theory of how the origin

---

* Owen's attacks on Darwin shifted wildly over the years. He could attack Darwin, on the one hand, for suggesting that humans were related to apes—which Owen never accepted—and, on the other hand, for taking too much credit for the development of evolutionary theory. Darwin found the depth of Owen's animosity perplexing. Jealousy at being shut out of Darwin's circles of intimates has been suggested as a motivation.

of life could come about, and would still be considered a reasonably good guess more than a hundred years later.

Darwin was still skeptical that such an act of spontaneous generation could take place in the Earth's current highly developed state. The problem was his own law of natural selection. The first organism would be, by definition, poorly adapted to survive. "At the present day such matter would be instantly devoured, or absorbed, which would not have been the case before living creatures were formed," he wrote. As time went on, Darwin became more open to the idea of spontaneous generation in the modern environment. Others among the growing community of scientists called Darwinists were even more confident that this untapped piece of evolutionary theory, the "vital spark," could be found, even that it could be recapitulated in a laboratory.

NOT LONG AFTER the publication of *Origin*, Darwin came under fire for failing to acknowledge his predecessors in the development of evolutionary theory. Darwin recognized the flaw almost immediately after publication, and he went about forming a list of those he should have acknowledged. His list eventually grew to include ten names, including that of his grandfather Erasmus, Lamarck, Wallace, and Aristotle, the last of which was an error by Darwin, who had mistakenly taken Aristotle's recitations of the views of others—which Aristotle did not share—as the Greek's own.

Today, the fact that the idea of evolutionary change was not first noted by Darwin has largely become irrelevant. He has become the de facto face of evolution, and the man behind the little fish with legs that people put on the backs of their cars. With the possible exception of Albert Einstein, no other figure is so closely identified with a scientific theory than Darwin is with evolution. And few besides Einstein come close in name recognition. Yet, while few really understand the principle of relativity beyond its iconic mathematical symbolization, $E = mc^2$, evolution's fundamental tenets are easily grasped.

Darwin transformed the way we view life. Considering the scope of the change, he did it in a remarkably short period of time. The rapidity with which our perspective changed is owed partly to the fact that England's liberalized society was prepared for the new theory, and partly to the general state of the life sciences, which were advanced enough that the mounting evidence of evolution was too hard to deny. But Darwin was ideally suited for the task of explaining evolution. His caution and hesitation, even what some supporters saw as cowardice, was a useful tool in winning over a skeptical public and doubtful scientists. He did not try to go too far, too fast.

If Alfred Russel Wallace had not first mailed his essay to Darwin, Wallace himself might have been recognized as the "discoverer" of natural selection and thus come to occupy Darwin's place in the pantheon of science. But Wallace may have been ill suited for such a role. Darwin was the more accomplished scientist and writer, and close to the organs of power and knowledge in the most powerful and influential state in the world, the United Kingdom. Wallace was a man of no means. He had no ties to universities, no friends in influential places in the sciences, and not even a university degree. In his later years, he flirted with disreputable forms of spiritualism and mesmerism. Under his leadership, acceptance of the basic tenets of evolution might have been an even slower process.

Yet Wallace might have been more daring. He was much more willing to tackle, from the beginning, thorny issues like the subject of human beings' evolution from apes. Darwin's critics saw such an evolutionary implication from the day *Origin* first appeared. So did the most sympathetic of his scientific contemporaries. It took Darwin thirteen years to thoroughly address the question in his book *The Descent of Man*.

Still, Darwin did address it publicly. That was something he never attempted in any significant way with the question of the origin of life. Despite his reluctance to face the subject head-on, his impact was profound. Before Darwin presented his model of evolution, it was not really even a single question. People asked where the first monkey came from, or the first shark. Earlier naturalistic explanations of the origin of life—from the Greeks to Buffon—had revolved around the question of the first of each species, but not the first *species*, period. Those who believed in spontaneous generation as the source of creation, like d'Holbach, believed that the embryo of an elephant or even a human being could be the result of spontaneous generation from inorganic sources. This view was still held by some scientists in Darwin's time. Lamarck's concept was essentially a version of this. But after Darwin, the question crystallized. It became the search for a *single* living organism to which all other living things were related. Those who once wondered about the first of each species now wondered about a single first ancestor of all of them.

In Darwin's growing brood of evolutionary thinkers, there were others willing to meet the question of the origin of life head-on. Over the last few decades of his life, Darwin would watch from the sidelines as one of his most promising disciples would engage in a famous struggle over the origin of life, pitted against a scientist who saw evolution, at least at first, as an enemy. By the time it was all over, the winner of the argument would achieve a legendary reputation that would rival Darwin's own. The loser would find himself all but forgotten.

PLEASANT, THOUGH THEY
BE DECEITFUL DREAMS

To explain all nature is too difficult a task for any one man or even for any one age. 'Tis much better to do a little with certainty, and leave the rest for others that come after you, than to explain all things.

—SIR ISAAC NEWTON, *Principia*, 1687

O N APRIL 7, 1864, a huge crowd filled the grand amphitheater at the Sorbonne in Paris's Latin Quarter. Among them were some of the cream of Parisian high society, including Princess Mathilde (the niece of Napoleon Bonaparte and cousin of the Russian tsar) and Amantine-Lucile-Aurore Dupin, who had become a famous writer under her pen name, George Sand, and was the lover of Chopin, as well as, it was whispered, the actress Marie Doval. Dupin was a regular at such events, always recognizable because of her taste for dressing in men's clothing. Sitting prominently in the front row was another writer known to all of Paris, Alexandre Dumas, the author of *The Count of Monte Cristo* and *The Three Musketeers*.

The lecture was part of a new series of free biweekly lectures hosted by the University of Paris, part of the university's push to generate public interest and support for its work. The Monday lectures were devoted entirely to science. Billed as "scientific soirées," they proved a huge draw. Part of the attraction was the array of technological devices the university had acquired for the amphitheater. Gas lighting could be raised or lowered at the touch of a button. An electric arc lamp, invented by Humphry Davy, could be used to project a single beam of light onto the stage, illuminating the speaker or selected exhibits. To the amazement of audiences, it could also be used to project photographic images that had been encased in glass slides.

The week before, six thousand people had shown up for the inaugural science lecture on the subject of the three physical states of matter delivered by the physicist Jules Jasmin. Most of the throng had been forced to stand outside, crowding around doors trying to get a peek at the goings-on inside. This week the soirée organizers expected an even larger turnout, owing to the popularity of the lecturer, a handsome and charismatic man with a reputation for giving captivating speeches, who was fast becoming the darling of the French scientific establishment. His name was Louis Pasteur.

The Sorbonne event represented a victory lap for Pasteur. For the previous three years, he had been engaged in a public debate over the topic of spontaneous generation with the man who had been considered the country's leading advocate of the theory, the well-respected naturalist Félix Pouchet, director of the Natural History Museum in Rouen. Their debate had generated so much attention that the French Academy of Sciences had decided to try to bring it to a resolution by offering the organization's prestigious Alhumbert Prize and a 2,500-franc award to the scientist who could shed the most light on the question. Pasteur had been judged the winner, and the academy had lauded his mastery of the experimental method. The victory had made him a hero in Catholic circles, where he came to be seen as the defender of traditional religious beliefs against the heresies of radical scientific materialism.

The French translation of Charles Darwin's *On the Origin of Species* had been in print for three years, and Pasteur began his address by summarizing the questions that were on the minds of these Parisians: whether "creation ought to be dated thousands of years or thousands of centuries past, whether species are fixed, or rather undergo a slow progressive transformation into new species." Pasteur was always a masterful orator, but never more so than that evening at the Sorbonne. Like Shakespeare's Marc Antony, who came to bury Caesar, not to praise him, Pasteur began by saying that he could answer none of those questions. He then went on to imply that he had answered them all.

Pasteur summarized the evolutionary theory by repeating a rough account of the concept from a book many in the audience would have read: *La mer* (*The Sea*), by the radical historian-turned-evolutionist Jules Michelet: "We simply take a drop of sea-water, and out of this water, which contains a bit of inanimate nitric matter, sea-mucus, or, as he calls it, fertile jelly, the first creatures emerge by spontaneous generation, transforming themselves bit by bit, they climb the ranks of creation, reaching, after, say, ten thousand years, the level of insects, and doubtless, after a hundred thousand years, the level of apes, and of Man himself." Behind it all, continued Pasteur, lay one question at the root of the whole evolutionary proposition: "Mightn't matter, perhaps, organize itself? Or . . . mightn't creatures enter the world without parents, without forebears?" If life were merely an outcome of natural processes, Pasteur told his audience, then they could arrive at no other conclusion than that "God is useless."

Pasteur then proceeded to recite the history of beliefs on spontaneous generation, from van Helmont to Needham and Buffon to the more modern scientists who embraced the idea. But there was something that all of these esteemed naturalists had missed, he said. At that point, the gas lighting dimmed until the room was dark except for a single beam of light dramatically projected onto the stage. There, in the illumination, he directed the audience's attention to the thousands of tiny dust particles that now flickered in the lighted haze above the stage. There in the dust, said Pasteur, was the reason so many great minds of the past had fallen into falsely believing they had witnessed the spontaneous generation of life: tiny microorganisms, invisible and countless, drifting through the air we breathe. Drawing upon the name used by preformationists, he called them "germs." The idea would eventually help make Pasteur one of the most legendary scientists in history.

———————

IN PASTEUR'S LATER YEARS, after his work on understanding the causes and prevention of infectious diseases had led to remarkable advances in medicine that would establish his place in the pantheon of the world's greatest scientists, he maintained that spontaneous generation was the most important question he had ever put his mind to. It was a question loaded with metaphysical importance, especially in France. As a crucial part of the broad evolutionary framework of Jean-Baptiste Lamarck, spontaneous generation was a theory that openly attacked the whole concept of a living world crafted by the hand of a deity. As France drifted ever deeper into politicized Catholicism, it also became an explicitly political question, one whose fortunes tended to ebb and flow with the notoriously fickle winds of postrevolutionary France.

Lamarck developed his theory at the turn of the nineteenth century, during the turbulent days of the revolution, while working as a professor at the Jardin des Plantes, as the Jardin du Roi had been renamed by the revolutionaries. He had been hired by Buffon himself and was widely recognized as Buffon's protégé. The young men of the revolution flocked to Lamarck's lectures about a natural world that was ever in flux and constantly evolving. He even dared to suggest that the natural world had the power to create life itself. Making such a claim would have once meant risking persecution or death, and it was a view held only by atheists on the fringes of science. For the time being, though, the revolution had made such ideas palatable.

As the French state and society grew more and more conservative with the restoration of the Bourbon monarchy, Lamarck found his notions increasingly attacked or discarded. By the 1830s, the stage had been set for a decisive showdown in France over evolution. It would be waged by two of France's most admired scientists, who had come to represent the two opposing poles of the evolutionary spectrum: Georges Cuvier and Étienne Geoffroy Saint-Hilaire.

Both, like Lamarck, had once held professorships in anatomy at the Jardin des Plantes. Both had also been close to Napoleon. Cuvier was a trusted member of the emperor's inner circle, charged with overseeing the education of Napoleon's son. Geoffroy was handpicked to accompany Napoleon on the short-lived French invasion of Egypt.

Their paths began to diverge after the Bourbon restoration, scientifically as well as politically. After Napoleon's exile, Cuvier—sociable, charismatic, and doctrinally flexible—had little trouble insinuating himself into the new order. He continued to rise through the ranks of the French scientific establishment, receiving a series of prestigious appointments, a royal peerage, and even an invitation to the coronation of Charles X in 1827.

Cuvier's own interest in naturalism stemmed from reading Buffon's *Natural History* as a boy. Soon, he had developed a fascination with paleontology that would make him one of the giants of French science. As far back as Leonardo da Vinci's time, natural philosophers had recognized that fossils were the preserved remains of creatures that had lived in the past. Buffon was one of the first to argue that these were often the remains of species that also no longer existed in the present world, but his conclusions were hard for most to accept. Why, they wondered, would God have created species only to let them die out? But without accepting the phenomenon of extinction, geologists struggled to explain the variations of fossil species they continued to find in the natural record every day.

Working with skeletons of Siberian mammoths, Cuvier became convinced that he was seeing entirely different species from those that walked the Earth in his day. He became certain after working with the fossilized bones of a creature he named a mastodon. His 1799 paper on the subject, considered a masterpiece in paleontology, established incontrovertibly that natural extinctions had indeed occurred in the history of the Earth. But Cuvier never accepted Lamarck's concept that new species were being formed, taking the place of those that disappeared. These abrupt changes in the fossils of particular regions from different periods of history were, for Cuvier, simply the result of migrations of already-existing species into the territories once occupied by other, now-extinct animals. Cuvier liked to point out that his own conception of natural catastrophes was remarkably consistent with those described in the Bible. Privately, Cuvier had little time for religion, and those who had known him in his younger days were surprised to find him resorting to charging his opponents with "materialism" to win scientific arguments. But he was ever the political opportunist.

Lamarck was the opposite. Doctrinally inflexible and socially awkward, he increasingly found himself isolated and ostracized, even by his old colleagues from the Jardin des Plantes. He spent his final years hopelessly in debt, struggling to find an audience for his revolutionary theories. When Lamarck died penniless and nearly blind in 1829, Cuvier penned a damning elegy for the French Academy of Sciences, full of left-handed compliments, such as praising Lamarck for being "gifted . . . with a lofty imagination."

Geoffroy preferred to lie low as the political winds shifted, throwing himself into his work. He had come to adopt a different, more Lamarckian interpretation of the fossil record than Cuvier did. Where Cuvier saw differences, Geoffroy saw similarities. A human hand, a bird wing, and a whale fin were all nearly identical in their underlying bone structure. They had simply been put to different use. What variances they possessed were acquired through centuries of Lamarckian adaptation. Geoffroy didn't go so far as to claim that all organisms evolved from a single original species, but he did argue that all vertebrates—every creature that had a spinal column—traced its history to a common ancestor. Cartoonists began to humorously portray Geoffroy as an ape. Evolutionists would often be portrayed this way in the press for the rest of the century.

Eventually, Cuvier and Geoffroy found themselves running separate wings in the Paris Museum of Natural History, where they used their authority to advance their clashing interpretations of the fossils and bones in their charge. In 1830, the French Academy of Sciences decided to stage a public debate between the two men, to be held at the museum. It ended up lasting nearly two months. Honoré de Balzac, Gustave Flaubert, and George Sand all weighed in with their own presentations at one point or another. By the time it was all over, it seemed as if all of France was following the proceedings. So, too, were many abroad. From Switzerland, the German writer Johann Wolfgang von Goethe followed every twist and turn with interest. In German academic circles, the concept of a wholly

naturalistic origin of life was more firmly held, and spontaneous generation was seldom in dispute. To a friend, Goethe likened the debate to the
eruption of a volcano. When the conservative French Academy of Sciences
finally declared Cuvier the winner, Goethe was disgusted. He attributed
the outcome to nothing more than an assertion of the old guard's power.
As France grew ever more conservative, support there for spontaneous
generation waned for the next two decades.

ONLY TEN YEARS OLD at the time of Cuvier and Geoffroy's famous clash, Pasteur grew up in the politically polarized world of French natural philosophy. It would color his view of science and his own place in it. Politically
conservative, Pasteur pitted himself against materialism when it came to
the life sciences. Like Cuvier, he stood for tradition and Catholicism.

Pasteur's rise to prominence in the world of French science had been
quick, but it was not easy or predictable. He was born in 1822, the son of a
poor leather tanner from the small city of Dole in the south. In college, he
struggled with chemistry and spent little time working in the life sciences
that would dominate his later career. His first university appointment was
as a professor of physics at the University of Strasbourg.

Pasteur had an early interest in the nature of crystals, a fascination he
shared with Andrew Crosse. At Strasbourg, the study of crystals led Pasteur to confront a puzzle about the nature of life for the first time. Working
with a crude polarimeter, a device that measures the ways chemicals inter-
act with light beams, Pasteur set about studying the effects of light when
passed through solutions of tartaric acid, a substance commonly found
in fruits like grapes and tamarind. Tartaric acid was also made synthetically and sold as a baking agent called racemic acid. Pasteur noticed that
tartaric acid and racemic acid, while compositionally identical, had very
different effects when he passed polarized light through solutions of them.
Tartaric acid neatly rotated the light clockwise, while racemic acid seemed
to have no effect on the light at all.

Pasteur was puzzled by this difference, but very close inspection with
a magnifying lens revealed that a solution of racemic acid that he'd left to

crystallize in a beaker on a windowsill had produced two types of crystals, seemingly identical in every respect but actually mirror images of each other. Pasteur painstakingly set about separating the two types of crystals into two small piles. When he dissolved the two piles separately, he found that each solution rotated polarized light beams in opposite directions, and that one of the crystal types was identical to natural tartaric acid. Pasteur realized that the synthetic racemic acid was a one-to-one mixture of the natural form of tartaric acid and its mirror-image isomer.

Pasteur had stumbled upon the phenomenon of chirality. Many molecules are what a chemist would call "chiral," meaning that they come in left- and right-handed forms that are mirror images of each other, like left- and right-handed gloves. These two forms have the same number of atoms, and the atoms are connected in the same way, but the two forms cannot be superimposed on top of each other, just as a right hand cannot be superimposed on a left hand. Most important biological molecules—such as the components of proteins or nucleic acids—are in fact *homochiral*, meaning that the orientation of every molecule is exactly the same, either right-handed or left-handed. The reason for the phenomenon of homochirality in biology would be a topic of debate among biochemists for another century. For Pasteur, the discovery reinforced his notions of a vitalist line separating the worlds of living and nonliving matter.

In 1854, PASTEUR WAS NAMED professor of chemistry and dean of the Faculty of Sciences at the University of Lille. Soon, he found himself turning to a subject that had even more obvious biological implications: the problem of alcohol. More precisely, how did the process of fermentation work to make alcohol? It was a mystery that had fascinated countless observers since ancient times.

Lille was in the heart of French sugar beet country. Back in the Napoleonic era, sugar had been hard to come by in France, owing to the decade-long blockade by Great Britain, so Napoleon, ever the micromanager, had encouraged the growth of a domestic sugar beet industry. French distillers had tried their hand at concocting a kind of rum from beet juice, but making it palatable was not easy. Its intense acidity was off-putting, and it had an unpleasant smell. By the time Pasteur arrived in Lille, beet liquor distillers were desperate. The father of one of his students, a wealthy industrialist, came to Pasteur to ask if he might look into the problems the local industry was facing. Though he had little experience with biology, Pasteur threw himself into finding a solution. Soon, his wife was writing her father that Pasteur was "up to his neck in beet juice."

In the cellar of an old sugar factory, Pasteur assembled a makeshift lab equipped with little more than a primitive coal-fired incubator. But he also used something that was surprisingly rare in a chemistry lab of the period: a standard student microscope that he borrowed from the university. Chemists of the time measured and weighed and experimented to arrive at abstract chemical formulas. Biologists and physiologists were the observers, using microscopes to explore the fine structure of their subjects. By actually observing fermentation in progress, Pasteur began to understand it in ways most of his fellow chemists did not.

Everyone knew that the key to making alcohol was the process of fermentation. However, not much concrete was known about what was actually happening during the process. Brewers and vintners knew about the presence of yeast in ferments. Antonie van Leeuwenhoek had been the first to see yeast cells, microscopic ovals that floated aimlessly under the lens of his microscope. But since they didn't appear to move by their own power, he did not consider them to be living organisms. The very word "yeast" was simply a derivation of the Old English word for "foam." By the time Pasteur began to contemplate fermentation, few scientists suspected that yeast cells, even if alive, were anything more than bystanders in the fermentation process. Antoine Lavoisier, a towering figure in the early history of chemistry, had practically ignored the presence of yeast entirely in his own attempt to describe fermentation. But in the late 1830s, a handful of scientists started to suspect that yeast were actually masses of microorganisms, and that they were key to fermentation.

Using his polarimeter, Pasteur saw that the ferment itself showed signs of asymmetry. From his study of quartz crystals, Pasteur realized that asymmetry meant the ferment was the product of living organisms, since only they could produce chemicals with purely asymmetric molecules. As the process went along, he also noticed that the yeast cells were changing shape, growing from little ovals into longer ones, then pinching off and dividing into themselves. He began to understand that they were living and, in fact, the *source* of the alcohol produced during fermentation.

Pasteur's hypothesis had all kinds of ramifications beyond distillation. Fermentation was not just a key to making alcohol; it was also suspected to be the chemical process that led to decay in all organic matter. By tracing the root of spoiling to microorganisms, Pasteur had opened the door to a new way to preserve food. In early 1862, another French scientist, Claude Bernard, showed that milk, wine, and beer could be boiled to kill off bacteria and mold, preventing spoilage and drastically increasing the amount of time they could be safely stored. The process came to be known as "pasteurization."

BY THE TIME Pasteur delivered his address on spontaneous generation at the Sorbonne, the discovery of pasteurization was nearly a year old. Yet Pasteur did not even see fit to mention it. To him, it paled in comparison to the metaphysical importance the French placed on spontaneous generation.

He first turned to the question of spontaneous generation in 1859, the year Darwin's *Origin* made its debut in England. Pasteur wrote to a colleague that he was going to "take a decisive step by resolving the famous question beyond the shadow of a doubt." The year before, Félix Pouchet had submitted a paper to the French Academy of Sciences claiming that experiments he had conducted proved the existence of spontaneous generation. Pouchet was a brilliant physician and naturalist, and the author of several well-received books on the natural philosophy of Aristotle. He was also a devout Catholic, much more so than Pasteur, who rarely attended church. Pouchet took pains to say that he was not challenging the idea of Christian creation as the original source of life, but merely trying to revive the question in the Augustinian sense. The process, he argued, was just one of the many tools used by God to propagate life. Nonetheless, Pouchet came under fire from conservative scientists who charged him with reviving materialist ideas discredited by Cuvier in his supposed victory over Geoffroy.

Pouchet couldn't have picked a worse time to submit his paper. Catholicism was growing increasingly politicized and ever more powerful in France. Its interests had become synonymous with the interests of the state. In 1848, the country held its first election by direct popular vote. Napoleon Bonaparte's nephew, Louis-Napoleon Bonaparte, won an overwhelming victory on the backs of a campaign that appealed directly to Catholics, promising to restore Catholic political power, to elevate the place of church in society, and to support the pope in Rome. Three years later, he staged a coup d'état and proclaimed himself Emperor Napoleon III. Censorship was restored, and some six thousand political opponents were sent to prison, some to penal colonies in French colonies abroad. Victor Hugo, who had become a living embodiment of the French national conscience, publicly denounced the emperor as a traitor and fled the country for a self-imposed exile that would last the rest of Napoleon III's reign.

Pouchet also had to deal with the issue of how his ideas played into the publication of Darwin's *Origin*, which dredged up all the old polarization on the issue that had surrounded the debate between Cuvier and Geoffroy. Those tensions grew in 1861 with the first appearance of a French version of *Origin*. The book had been translated by Clémence Royer, a radical who would one day become the first woman appointed to the French Academy of Sciences. Royer's translation included her own preface containing a lengthy diatribe against the Catholic Church. Darwin, who knew nothing of Royer when he agreed to accept her as translator, wrote to his friend, the American botanist Asa Gray, that Royer was "one of the cleverest and oddest women in Europe" who "hates Christianity and declares that natural selection and the struggle for life will explain all morality, the nature of man, politiks." Darwin added that Royer planned to write her own book on these subjects, "and a strange production it will be."

In the early part of his career, Pasteur often picked fights with more established scientists as a way of enhancing his own prestige. Pouchet was sixty-two at the time they began to clash, and he was considered the country's leading expert on spontaneous generation. Pasteur was thirty-seven. Though he had impressed many with his work on fermentation, he had spent little of his career tackling biological questions.

In order to topple Pouchet's results, Pasteur looked for flaws in the older experimenter's methods. Part of the difficulty in studying spontaneous generation was the need to exclude air. At the time, people believed that all organisms needed oxygen to survive, so oxygen inevitably had to be reintroduced to create conditions in which spontaneous generation could occur. Yet his experience with fermentation led Pasteur to wonder whether microorganisms weren't simply being introduced through the air itself, carried by dust. To test his hypothesis, he devised a simple apparatus that would henceforth become an iconic symbol of Pasteur's genius: the swan-necked flask. As beautiful as it was functional, it was round at its base and had a long, thin neck, curved like the neck of a swan. Such a neck would allow air to pass, but not dust or, presumably, airborne microorganisms, which would find themselves trapped in the neck's many curves. He placed broth in his flasks, the same substance once used by John Needham, boiled it in place, and then sealed some flasks with a flame while leaving others exposed to the air. He let them sit for days, then weeks. Those left sealed remained sterile, but more significantly, those left open to the air did too. As a final coffin nail, he broke the seal on the ends of the swan necks of the closed flasks and left them open to the air. They, too, remained sterile.

A commission set up by the Academy of Sciences to judge the contest declared Pasteur the "decisive" winner, which paved the way for his triumphant lecture at the Sorbonne. In reality, the deck had been stacked against Pouchet from the outset. The commission was overwhelmingly composed of conservatives, many of whom had already stated their outright rejection of spontaneous generation on the grounds that it was a materialist doctrine. In their judgment, they wrote, Pasteur not only had demonstrated that "brute matter cannot organize itself in such a way as to form an animal or plant," but also had proved the theory of vitalism, that a "life force has been passed on successively through an uninterrupted chain of being

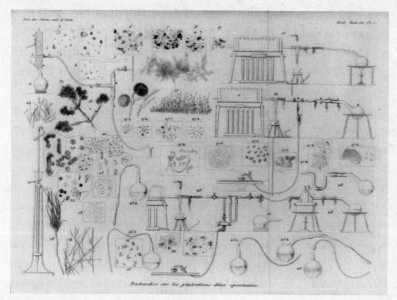

Pasteur's experiment on spontaneous generation.

since creation." The outcome secured Pasteur's reputation as a master of the experimental method, and his identification of microorganisms that traveled through the air would lead him closer to one of his most important scientific contributions, the germ theory of disease. But the question of spontaneous generation was not quite dead. A new debate was brewing just across the English Channel, where support for the Darwinist concept of evolution seemed to grow stronger by the day.

ON A WINTER DAY in 1864, eight distinguished-looking gentlemen gathered at the St. George's Hotel on Albemarle Street near London's Piccadilly Circus, just down the street from Burlington House, the huge, elaborate mansion that served as the headquarters of the Royal Society. The dinner party consisted of some of the brightest and most distinguished scientific minds in the British Empire, drawn from a wide range of disciplines. One, the anatomist Thomas Huxley, later claimed that among them, they could have managed to produce "most of the articles to a scientific Encyclopedia." That night, they decided to form an exclusive semiformal club to foster, as one of the attendees, Joseph Hooker, wrote to his close friend Charles Darwin, "devotion to science, pure and free, untrammeled by religious dogmas." They hoped to use their influence as a kind of behind-the-scenes steering committee that would usher the world of British science into the modern age. Three of them would go on to hold the title of president of the Royal Society, and five would hold the same position at the British Association for the Advancement of Science. They called themselves the X Club.

The men decided to keep their membership small. No more than ten would be allowed in. Eventually, they added a ninth member, but the tenth position was never filled. The position might as well have been held in absentia by Charles Darwin. Roughly half of the group was drawn from Darwin's inner circle, and the X Club quickly became an informal operational command center for advancing the evolutionist cause in Great Britain. One of the club's first acts was to ensure that Darwin was granted the Copley Medal, the highest honor of the Royal Society. The competition for the award pitted Darwin against Adam Sedgwick, who had been nominated by Richard Owen. Darwin barely edged out his old professor in a vote by the award committee, a highly symbolic moment that showed just how quickly the face of British science was changing. Darwin characteristically avoided the presentation ceremony, citing ill health.

Huxley had argued furiously on Darwin's behalf. He managed to win over many voters still skeptical of *Origin*'s message. Huxley himself had once been a skeptic of evolutionary theory. But over the previous decade, he had become one of its most zealous defenders. He called himself "Darwin's Bulldog." The nickname stuck.

Darwin was not a gifted speaker. He could easily become sidetracked when speaking in public, and he had a habit of overzealously waving his hands through the air in a way that made people uncomfortable. Huxley, on the other hand, was a skilled orator and filled that role when it was called for. The first and most famous time was in 1860, at a meeting of the

British Association for the Advancement of Science that has since assumed an almost mythical status in evolutionist history. It was held at Oxford, in the diocese and auspices of Bishop Samuel Wilberforce, an old-guard Tory whom the liberal *Daily Telegraph* once called a representative of those who never "advanced one iota beyond their ancient notions."

Wilberforce and Huxley joined in a debate, arranged by the conference organizers, over the merits of evolutionary theory. Seven hundred people showed up to watch the proceedings—so many that the event had to be moved to Oxford's Museum of Natural History, a grand hall designed in the Gothic Revival style, with an engraving of an angel hovering prominently above the entrance. At the end of his own presentation, Wilberforce, who had been coached by Richard Owen, asked Huxley whether his descent from an ape could be traced to his grandfather's or his grandmother's side. The audience, already tense over the gravity of the issue being discussed, reacted in loud rancor. Struggling to make himself heard over the din, Huxley replied that he would rather be descended from an ape than from one who impeded the cause of science. A woman fainted. In the pandemonium, Robert FitzRoy, Darwin's old captain aboard the HMS *Beagle*, stood up and raised an enormous Bible over his head, commanding the audience to return to God.*

It was the first time Christianity had ever been pitted in a formal debate against science. Huxley emerged from it as a hero not just to evolutionists, but to rationalists of many stripes, even among many liberal-minded Christian theologians, a group both he and Darwin hoped to cultivate. Clergymen actually made up a majority of the audience at the Oxford debate. Huxley understood that among them were many who could be won over to the evolutionary cause. Though he could be at times uncompromising, he often took more nuanced views toward religion, especially in his public comments. Wherever possible, Huxley tried to ensure that the

---

* Five years later, FitzRoy, beset by financial troubles, would commit suicide by cutting his throat with a razor.

secularist camp didn't go too far. He coined the term "agnostic" to describe his own materialistic lack of religious faith. It made him seem less threatening than the outright atheists.*

ONE OF THE MOST important tasks of the X Club was to build up a new class of professional British scientists. Natural philosophy had long been the exclusive purview of the rich and the privileged, and roughly half of the Royal Society's five hundred members had been selected on the basis of nothing more than their social standing. The first woman would not be admitted until 1945. The X Club advocated for a smaller institution composed of actual scientists drawn from all classes, rather than a bloated organization full of highborn men for whom science was merely a fanciful pastime. Men like Alfred Russel Wallace and Louis Pasteur had risen through the ranks of a scientific world that was increasingly becoming a meritocracy. So, too, had the self-educated Huxley.

The eighth child of a seldom-employed math teacher, Huxley had received a mere two years of formal schooling, the last of which ended when he was ten. Nonetheless, he had taught himself German, Latin, and Greek. Deeply in debt after putting himself through medical school, Huxley had signed on as an assistant surgeon in the Royal Navy, assigned to the HMS *Rattlesnake* for an exploratory voyage to Australia and New Guinea. His well-received observations on marine invertebrates convinced him to concentrate solely on naturalism.

Huxley knew firsthand how difficult it was to be a self-supporting scientist, and he relished the task of mentoring budding young evolutionists in Great Britain and abroad. He not only advised them on theory and their use of experiment, but also tried to instill his idea of scientific decorum and char-

---

* X Club members were instrumental in beginning two scientific publications—first the *Natural History Review* and then the *Weekly Reader*—to which the membership contributed most of the articles. Both struggled economically, but eventually a third publication stuck: the journal *Nature*, published by Alexander Macmillan, a friend of Huxley's who was famous for his "tobacco parties," which were like an ongoing book club centered around Darwinism and other aspects of evolutionary theory. Huxley penned the first article in the inaugural issue, and other X Club members contributed heavily. With a readership of nearly 450,000, *Nature* is today perhaps the most influential scientific periodical in the world.

acter. Charged with guiding evolution through the treacherous waters of theology and entrenched scientific thinking, Huxley, who was usually known for his combativeness, often preached caution and the value of knowing when to retreat from unsupportable assertions. When he did, he often used the phrase "eat your leek," meaning "take your medicine." It wasn't long before he had a brilliant young scientist on his hands who would not take that bit of advice when it came to the subject of the origin of life, a physician named Henry Charlton Bastian. Before he was through, Bastian would become the first of a long line of Darwinists to try to experimentally prove the critical moment that Darwin himself had shied away from: the origin of life. Bastian would also become the last serious scientist to try to prove the old doctrine of spontaneous generation, at least in the way spontaneous generation had been conceived of since Aristotle, as a common ongoing occurrence.

THE MIDDLE-CLASS son of a simple merchant from Cornwall, Henry Charlton Bastian was part of that growing group of English scientists from modest backgrounds whom Huxley was predisposed to support. A prodigy, Bastian had authored a sophisticated book on botany by the time he was nineteen. Before graduating at the age of twenty-three from University College London, taking four degrees, he had published an important zoological study that revealed one hundred new species of nematoid worms.

Upon graduation, Bastian's first position was assistant conservator of the university's Museum of Morbid Anatomy. Soon after, he accepted a medical residency at the newly opened Broadmoor Criminal Lunatic Asylum, England's first prison for the criminally insane. Both were jobs most people would have found distressing, but they were right up Bastian's alley. His specialty was neuroscience, and he was deeply interested in the mystery of consciousness. He came at it from a perspective called physicalism. In the nineteenth century, most people understood consciousness to be ethereal. Physicalists, by contrast, maintained that consciousness was a purely physiological construct. It was controlled by a body organ, the brain, just as the heart controlled the flow of blood. All components of a person's ability to think could be traced to corresponding areas of the brain, such that injuries in certain brain areas could result in changes in mental functioning and even personality.

Guided by this concept of the material basis of thought, Bastian made huge strides in our understanding of aphasia, a condition that follows strokes. By 1868, his work had earned him membership in the Royal Society, a sterling reputation in the world of British medicine, and an appointment to the National Hospital, Britain's first hospital to specialize in neurological disorders. Previously, people with brain injuries such as aphasia had often been confined to insane asylums.

At the National Hospital, Bastian began to undertake a series of experiments hoping to reopen the door to spontaneous generation, a door that had, for many, been authoritatively closed by Pasteur. Behind a privacy screen in an examination room, Bastian set up a lab and began conducting experiments on boiled hay infusions that had been hermetically sealed. Like John Needham's before him, Bastian's experiments seemed to support the occurrence of spontaneous generation.

Bastian believed he was simply forging ahead to answer the question that Darwin had left woefully unresolved: how the whole unfolding evolutionary procession had begun. Spontaneous generation had always been a key part of Bastian's concept of evolution, a concept he learned about from the same man who had first instructed Charles Darwin: the radical professor Robert Grant, who had moved on from Edinburgh to University College London, an institution founded on the principle of secular education. There, Grant continued to preach evolution in the broad Lamarckian sense, which included an explanation for the origin of life. To stress spontaneous generation as not simply a source of life in the old Aristotelian sense, but the source of all life in the evolutionary sense, Bastian began using a different word to describe the phenomenon. He called it "archebiosis," Greek for "the beginning of life."

The medical establishment embraced Bastian's efforts to revive the question of spontaneous generation. Medicine then had a reputation as the most radical field in British science, one that still welcomed the ideas of men like Robert Grant. Many of Bastian's initial writings on spontaneous generation were published in Britain's premier medical publications. In the pages of the *Lancet* and the *British Medical Journal*, Bastian argued against there being any impassable boundary between the living and nonliving worlds. Nature did not work in one way through the long history of evolution only to allow for some kind of irreproducible miraculous event at its beginning. To Bastian, evolution was a continuous process that had begun before any living organism appeared on the planet Earth. Combined with his experimental results, the articles would form the basis of his first book on the subject, *The Beginnings of Life*.

Many evolutionists were thrilled that someone was filling the gap Darwin had deliberately left open. Alfred Russel Wallace wrote a glowing review, and Darwin read Bastian's book at Wallace's recommendation. In a letter to Wallace, Darwin called Bastian's general argument that organic matter must have been formed out of inorganic "wonderfully strong." Still, he remained skeptical of Bastian's claim that there was compelling experimental evidence in favor of spontaneous generation, although, he added, "I should like to live to see archebiosis proved true, for it would be a discovery of transcendent importance."

Though his book was initially well received, Bastian and the question of spontaneous generation were about to become perilously linked to another argument that was gaining steam in the scientific community. It revolved around the question of the nature of disease.

In Great Britain, it was a question of some urgency. For the previous four decades, Britain had been devastated by deadly outbreaks of cholera morbus, a condition first encountered by British soldiers in India in 1817. By 1831, the disease had spread throughout Russia and from there on a British ship to the English port of Sunderland, where local officials had ignored orders for a quarantine. The first outbreak killed more than fifty thousand people. By the time Bastian began working at the National Hospital, successive waves of the disease had claimed more than 250,000 lives. On the streets, people started calling it the "blue death," after the color that would appear on the faces of many of its victims.

A century later, cholera would be known as a relatively benign condition easily treated by intense hydration and replacement of mineral salts. In Bastian's time, the typical treatment for such diuretic conditions was to actually *restrict* the amount of water that those afflicted drank, which usually made the disease fatal.

Cholera is caused by a bacterium now known as *Vibrio cholerae*, but in the nineteenth century, most British medical men believed that illnesses were passed through the air as vapors. These were called "miasmas," from the Greek for "pollution," and the theory came to be called the miasmatic theory of disease. Medical vernacular is still filled with words suggesting an airborne source of illness. "Malaria," for instance, means "bad air." In London, where industrialization and its concomitant burning of coal had led to the persistent grayish-brown "London fog"—what we would now

Burning barrels of tar to ward off miasmas during the Manchester cholera outbreak of 1832.

call smog—the idea that air could be bad for one's health had a particular salience. So did the fact that disease spread rapidly among the growing ranks of the urban poor, who lived in filthy, overcrowded slums without proper sewage. To combat cholera, people took steps to contain the unhealthy vapors that had become associated with its spread. Barrels of tar and vinegar were set afire in infested streets to purge the air. Houses were doused with solutions made from lime.

By the time of Bastian's investigations into spontaneous generation, a competing theory had started to gain traction. The zymotic theory, now more commonly known as the germ theory of disease, held that many illnesses were caused by microscopic organisms. It was not a new theory, but the biggest hurdle it faced was a simple fact that almost everyone who dealt with disease understood: one did not actually have to touch an infected person to catch a disease like cholera, but could merely be in the presence of an infected person, breathing the same air. The means of transmission in such cases was unknown but was about to be solved by Louis Pasteur.

In 1865, PASTEUR RECEIVED a letter from Jean-Baptiste Dumas, a famous chemist and devout Catholic supporter of Napoleon III. From the south of France, the heart of France's growing silk industry, Dumas wrote that, "misery is greater here than anything one can imagine."

Since the late eighteenth century, France had been steadily encroaching on the Chinese monopoly of silk production. Most of the silk production was centered around the city of Lyon, which had become the silk capital of Europe. Whole forests had been cut down to make room for the golden-leaved mulberry trees that supplied the leaves upon which the silkworms fed. But a mysterious disease spreading among the silkworms had begun to cripple the economy of the region. In desperation, the silk producers turned to Pasteur. Though he had little experience in biology outside his work on fermentation, Pasteur devoted himself to finding the source of the disease, which he eventually traced to parasitic microbes that preyed on silkworm eggs.

Pasteur's discovery saved the silk industry. It also led him to realize that he could turn his expertise to the problem of infectious disease in people, where he would make his most lasting impact as a scientist. Pasteur's interest in disease had a personal dimension: he had lost two young daughters to typhoid fever.

Soon, Pasteur brought his experience with the spontaneous generation question to bear on the notion that germs were the cause of most disease. If his theory of airborne bacteria was indeed true, it stood to reason that bacteria could spread disease in the same way. The transmission of the disease by bacteria would solve germ theory's biggest conundrum, the mystery of infection without direct contact.

Initially, the idea of airborne bacteria causing disease was a notion on the fringes of the medical community, particularly in Britain. Doctors could often see bacteria in samples taken from infected patients, but they usually attributed their presence as a side effect—a *result* of the disease—rather than the cause. Most British physicians believed these bacteria were spontaneously generated, and they doubted Pasteur's assertion that germs could travel through the air. Bastian became a champion of miasmatic theory and argued that his own experiments on spontaneous generation provided a more sensible explanation for the presence of bacteria. Rather than floating through the air, they were spontaneously generated results of infection. The conflict over spontaneous generation thus became wrapped up in the conflict between the zymotic and miasmatic theories of disease.

AFTER HUXLEY, the next-most-important member of the X Club was the brilliant physicist John Tyndall, holder of England's singularly most prestigious scientific post, Professor of Natural Philosophy of the Royal Institution, where he had succeeded the great Michael Faraday. Tyndall built a sterling reputation through his experimental work on the electromagnetic properties of crystals. Later, he made huge strides in explaining the effects of infrared radiation on the atmosphere and the composition of ozone.

In his spare time, Tyndall was an avid mountain climber. He was the first person to climb the Weisshorn, one of the tallest peaks in the Swiss Alps, and he was one of the first to climb the Matterhorn. During an alpine expedition in 1869, Tyndall slipped in a rocky pool, severely cutting his leg on a granite outcropping. The resulting abscess had nearly killed him. Tyndall became convinced that airborne bacteria were responsible for his brush with death. He became a leading advocate of Pasteur's theory that germs carried through the air were the cause of disease.

Tyndall's advocacy of germ theory put him at odds with Bastian. Soon, the two men were facing off over germ theory and spontaneous generation in a series of letters to the editor published in the *Times*. Tyndall saw this as a battle to encourage the idea of a professional scientist as opposed to the "quackery" he railed against in the medical community. Bastian meanwhile became the champion of physicians, defending them against encroachment by interlopers from other branches of science who had no experience in medicine.

This was the beginning of Bastian's fall from grace in the eyes of the X Club. Huxley was appalled to see the argument between Tyndall and Bastian being waged in the *Times*. Huxley was striving to maintain a united front capable of advancing the evolutionist cause into mainstream Britain, and he believed it dangerous for the scientists he counted as allies to turn on each other in public forums, especially when one of those forums was the most widely read newspaper in Britain. What's more, Bastian was breaking Huxley's cardinal rule that scientists should show deference to their more experienced and accomplished colleagues.

Bastian had begun to try Huxley's patience even before coming into conflict with Tyndall. Early in Bastian's experimental work, Huxley had taken the young man under his wing. In Bastian, he spotted an opportunity to bring another promising young scientist into the web of influence of the X Club. He had personally watched Bastian conduct his experiments on several occasions to ensure the validity of Bastian's experimental evidence. But Huxley was concerned when traces of moss, considered too complex to be spontaneously generated, appeared in one of Bastian's supposedly hermetically sealed and sterile tubes. There was no accounting for their presence. To Huxley, this fact called into question the eventual appearance of microbes in Bastian's later experiments. As Bastian moved closer to publishing what Huxley knew would be a contentious book, Huxley advised Bastian to delay publication. On a subject as inflammatory as the origin of life, one had to proceed cautiously.

Huxley himself had learned that lesson the hard way. In 1868, he had been examining an old specimen of mud dredged from the Atlantic seafloor when he discovered a curious organic slime growing on its surface. The slime didn't resemble any known living organism and appeared to have emerged spontaneously in a sterilized sample. Huxley speculated it might be a kind of missing link between living and nonliving matter. He called the slime *Bathybius haeckelii*, in honor of the German philosopher and evolutionist Ernst Haeckel's idea that life had begun with a substance the German naturalist Lorenz Oken had named *Urschleim* ("primordial slime"). In his book *The History of Creation*, Haeckel had used Oken's slime to explain the first rung on his proposed evolutionary ladder. But other scientists were not convinced, seeing *Bathybius* as no more than a common fungus. Huxley sensed that he was precariously close to an Andrew Crosse moment. When *Bathybius* was finally shown to be nothing more than calcium sulfate precipitated from seawater by the ethanol used to preserve specimens, he promptly ate his leek, recanting in a letter published in *Nature* and a mea culpa delivered before the British Association for the Advancement of Science.

Huxley's error had been seized upon by opponents of evolution as proof that the whole evolutionary scheme was faulty. The episode taught Huxley the dangers of trying to move too fast, particularly when the evidence was shaky. He had no doubts that life had once arisen in the past, but the very distant past, when conditions on the

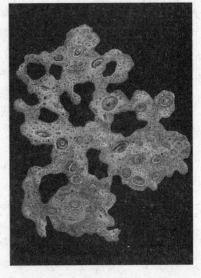

Drawing of *Bathybius haekelii* as seen under a microscope.

Earth were very different. He had moved to a position similar to Darwin's: that the conditions for spontaneous generation were no longer present in the current state of the natural world. Just as Bastian had adopted the name "archebiosis" to describe spontaneous generation, Huxley tried to breathe new life into the concept by giving it a new name. His choice of words was telling. He began calling the process "abiogenesis," Greek for "nonbiological creation." The term avoided the connotation of a process ongoing in the present, like the one Aristotle had once imagined. And unlike Bastian's term "archebiosis," the term "abiogenesis" left room for doubt that it was the source of *all* life. Huxley strongly believed that archebiosis was indeed the source of all life, but he felt it was enough simply to imply it. The idea didn't need to be shoved into people's faces.

BASTIAN'S BOOK, *The Beginnings of Life*, posed a challenge for Huxley. Rank-and-file evolutionists were inspired by Bastian's attempt to explain the beginning of evolution, but his anticreationist language frightened religious moderates, threatening to undermine Huxley's attempts to bring these moderates into the evolutionary fold. One of those who could see the overwhelming scientific evidence for Darwin's theory but was repulsed by its implications for his religious belief was the American mathematician George Barnard, the president of Columbia University in New York. Barnard had read Bastian's *Beginnings* and found himself swayed by its reasoning. Yet he still felt compelled to reject all of evolutionary theory because of what it meant to his spiritual beliefs. In an otherwise unremarkable paper that he authored about the germ theory of disease, Barnard meandered into his thoughts about the challenges that evolution and spontaneous generation posed to his own personal religious beliefs. It amounted to one of the most eloquent defenses of faith against reason ever written:

> We are told, indeed, that the acceptance of these [evolutionary] views
> need not shake our faith in the existence of an Almighty Creator. It is
> beautifully explained to us how they ought to give us more elevated
> and more worthy conceptions of the modes by which He works His

will in the visible creation. We learn that our complex organisms are none the less the work of His hands because they have been evolved by an infinite series of changes from microscopic forces of light and heat and attraction acting on brute mineral matter. . . . It is indeed a grand conception which regards the Deity as conducting the work of His creation by means of those all-pervading influences which we call the forces of nature; but it leaves us profoundly at a loss to explain the wisdom or the benevolence which brings every day into life such myriads of sentient and intelligent beings, only that they may perish on the morrow of their birth. But this is not all. If these doctrines are true, all talk of creation or methods of creation becomes absurdity; for just as certainly as they are true, God himself is impossible . . . if, in my study of nature, I find the belief forced upon me that my own conscious spirit . . . is but a mere vapor, which appeareth for a little time and then vanisheth away forever, that is a truth . . . for which I shall never thank the science which has taught it me. Much as I love truth in the abstract, I love my hope of immortality still more; and if the final outcome of all the boasted discoveries of modern science is to disclose to men that they are more evanescent than the shadow of the swallow's wing upon the lake . . . give me then, I pray, no more science. Let me live on, in my simple ignorance, as my fathers lived before me, and when I shall at length be summoned to my final repose, let me still be able to fold the drapery of my couch about me, and lie down to pleasant, even though they be deceitful dreams.

As Tyndall and Bastian continued to debate the merits of germ theory, their argument increasingly revolved around the question of spontaneous generation. The important point was whether Pasteur had, in fact, proved the existence of airborne germs. Tyndall began trying to undermine Bastian's results by seizing upon Huxley's low appraisal of Bastian as an experimenter. The argument became personal for Tyndall. In his eyes, what Huxley saw as poor experimental standards were a deliberate misrepresentation of results. Pasteur, goaded on by Tyndall, threw the weight of his reputation at Bastian. A July 1877 letter from Pasteur to Bastian revealed not only the depth of Pasteur's passion on the issue, but the combative spirit that belied the Frenchman's well-cultivated benevolent image: "Do you know why it is so important to me to fight and defeat you? It is because you are one of the main adherents to a medical doctrine that I consider extremely harmful to the art of healing, the doctrine of the spontaneity of all disease."

Huxley largely stayed out of the fray, but many of his subordinates joined in with relish. Some of the most aggressive criticism of Bastian came from the young zoologist Ray Lankester, who would go on to become one of the most influential evolutionists of the early twentieth century. In a vicious review of *The Beginnings of Life* in the *Quarterly Journal of Microscopical Science*, Lankester called Bastian a "mesmerized victim of delusions . . . such delusions form a very interesting psychological study, and it is only when we have obtained a proper conception of Dr. Bastian as an abnormal psychological phenomenon that we can hope rightly to appreciate the whole of statements made in his book."* The savagery of the attacks on Bastian that emerged from the Darwinist camp perplexed many in the larger world of evolutionary science, particularly those observing from other countries.

To put the argument to rest, Tyndall eventually decided to do something that was, as a physicist, a departure for him. He devised a biological experiment, one he hoped would discredit Bastian's work. He knew the key would be tackling the problem of airborne contamination, the same problem that had once led Pasteur to his swan-necked flask design. Tyndall's solution was to create an environment filled with what he called "optically pure air." He constructed a test-tube cabinet with a clear-glass viewing window. The sides and bottom of the cabinet were coated with a sticky syrup of glycerin. It acted like a modern-day lint roller, and eventually all the dust—and presumably Pasteur's airborne germs—stuck to the sides and bottom. In his cabinet, he tested different solutions for signs of spontaneous generation. But he didn't get the results he had expected. Some of

---

* In private, it was Lankester's sanity that worried Huxley. Huxley wrote to a friend that "there is what we call 'a screw loose' about him. I don't know exactly what screw it is, but there is something unstable about him" (Strick, *Sparks of Life*, 101). .

his tubes did, in fact, grow microorganisms. In trying to determine why, Tyndall hit upon something that seemed to unravel the long, contradictory history of research into spontaneous generation ever since van Leeuwenhoek's discovery of microbes. Tyndall identified heat-resistant spores that could survive intense temperatures, which came to be known as Cohn's spores after the German bacteriologist Ferdinand Cohn. Tyndall went on to devise a method for killing the spores by repeated heating, known as tyndallization, a process that is still commonly used.

Tyndall's discovery explained the appearance of microbial life in Bastian's test tubes: they were simply the offspring of heat-resistant spores that had survived Bastian's attempts at sterilization. The movement of germs through air came to be accepted by the medical community, as was the germ theory of disease. Bastian, who never could quite bring himself to eat his leek, saw his star descend in the scientific community where he had once showed such promise. He never wavered in his belief that spontaneous generation was a common, repeatable phenomenon, a belief that was almost universally abandoned by the scientific community in the wake of Tyndall's experiments. Though Bastian retained a great deal of respect among physicians, he was seldom again taken seriously in the larger world of science. The Royal Society refused to publish any more of his papers.

In Bastian's later years, when he turned to the private practice of medicine, he struggled financially as well, though he never quite reached the depths Lamarck had fallen to. Through it all, he continued to write books, well into the twentieth century, on his ideas regarding the origin of life. None were much more than a rehashing of his first. What little attention they garnered was not positive. An editorial in the *Lancet* that followed one of his last publications betrayed just how far he had fallen even in the eyes of the partisan world of medicine: "To our mind the position is quite unchanged, and we ourselves still remain unconvinced, save, of course, of the courage and good faith of Dr. Bastian."

Bastian represented the last gasp of the old Aristotelian idea that spontaneous generation was a normal, commonplace occurrence. But the idea that life could emerge from nonlife was not forgotten. With the exception of those who subscribed to the idea that life had been "breathed by the Creator into a few forms or one," for evolutionists, Huxley's concept of abiogenesis remained the only possible answer to the question of how life came to exist on Earth. Despite his own belief that abiogenesis was likely naturally possible only under the very different conditions that prevailed on the early Earth, Charles Darwin never lost hope that the process could be duplicated in a laboratory. In an 1882 letter to the Scottish geologist Daniel Mackintosh, Darwin wrote, "Though no evidence worth anything has as yet, in my opinion, been advanced in favour of a living being, being developed from inorganic matter, yet I cannot avoid believing the possibility of this will be proved some day."

Among the biggest obstacles in proving such a hypothesis was a lack of the most basic knowledge about the early planet. Scientists suspected that the Earth had been different when it was young, but they could not begin to appreciate just how great those differences might have been. Though the concepts of geological change espoused by geologists like Lyell seemed radical at the time, they did not even approach the truth of just how much the Earth's atmosphere and geological composition had changed across the eons. Nor did anyone understand just how long the Earth—and life on it—had been around. The answer would have shocked them.

NO VESTIGE OF A BEGINNING

Where wast thou when I laid the foundations of the earth?

—BOOK OF JOB, 34:4–7

THE PLANET EARTH was created on October 23, 4004 BC, sometime in the afternoon. That is, at least, what the majority of people in the West believed right up into the nineteenth century. Most might not have given an exact date. That was the work of Irish archbishop James Ussher, who, in his 1650 book *Annals of the Old Testament*, was able to nail it down precisely from the available references in the Bible. Nevertheless, most people believed the Earth was very young. They also believed, as Voltaire had believed, that the geography and geology of Earth had not changed much since the dawn of creation.

Yet, as many natural philosophers considered the Earth more carefully, and as human beings dug deeper beneath its surface, they began to realize that the geology of the Earth is like a history book. In canyons and mountains, in deserts and rivers, observers could see the long, slow processes of change that had formed the Earth's varied terrain. During his trip to the Nile, Herodotus speculated about the vast quantities of black sediment—the *kemet*—that had collected along the riverbanks. He figured it had taken many thousands of years for the delta to form. A thousand years after Herodotus, Arab scholars in the Iberian Peninsula and elsewhere wondered about the fossils of sea creatures in rocks found in the desert. These told them that water had once flowed over those long, arid stretches of sand, and vanished long, long ago.

By the late seventeenth century, geological clues had led European observers of the natural world to challenge the biblical account of creation. A Scottish geologist by the name of James Hutton, observing the slowness of geological change in the Scottish highlands, imagined the Earth was so old that it showed "no vestige of a beginning." In France, Buffon came up with an age based on scientific reasoning alone. His attempt grew out of a remarkable hypothesis of how the solar system came to be formed, a concept in some ways not too different from that held by modern astronomers. Planets, Buffon believed, were created by the collision of large, astral, rocky bodies. Our solar system was formed by a comet colliding with the sun, which would have left the Earth initially very hot. This event, he imagined, was followed by a long process of cooling that still hadn't ceased. Buffon rather ingeniously assumed he could measure the age of the Earth by figuring out just how long it took rocks to cool. By conducting experiments in which he heated and cooled iron spheres, Buffon eventually came up with an estimate of 74,832 years.* For his troubles, the Vatican threatened Buffon with excommunication. He contritely issued an apology and, with a wink, continued using the figure in his later writings.

Buffon's measurement was scientific, but it wasn't very accurate. By the early twentieth century, scientists envisioned an Earth that was billions of years old, so old that Buffon's estimate of about seventy-five thousand years would be much closer to Archbishop Ussher's than to the more modern estimates. This new understanding of how truly old the Earth was, and how much it likely had changed since its beginning, would have a huge impact among scientists who were still searching for answers to the great puzzle Darwin had left unsolved, that of the origin of life.

THE START OF THE twentieth century saw an explosion of scientific advancement in the Western world. In physics, Albert Einstein published his "quadfecta" of papers that, among other things, paved the way for the development of quantum mechanics and introduced the world to his spe-

---

* From those same experiments, Buffon estimated that in 93,291 years, the Earth would become too cold to support life.

cial theory of relativity. A Belgian Roman Catholic priest named Georges Lemaître proposed the Big Bang theory to explain the origin of the universe. The German chemist Fritz Haber devised an industrial synthesis of ammonia, leading to manufactured fertilizer that fueled an agricultural revolution and an unprecedented population explosion. The same chemical process was used to make the explosives that fueled the carnage of two world wars.

Yet despite remarkable technological progress in nearly every field of science, the subject of the origin of life had languished in the West. In the decades following the rejection of Henry Bastian's experiments on spontaneous generation, the question no longer attracted the greatest minds in science. School textbooks often presented the matter as having been closed by Louis Pasteur, whose experiments they said conclusively proved that life could never have sprung from nonlife. Most of the world's most important evolutionists recognized that abiogenesis had probably occurred sometime in the Earth's ancient past. But until the early twentieth century, they had very little idea what that ancient Earth looked like, or even how ancient it was.

In 1933, Sir Frederick Hopkins, a Nobel Prize–winning biochemist and president of the Royal Society, summed up the dire state of the research in a speech before the British Association for the Advancement of Science. "Though speculations concerning the origin of life have given intellectual pleasure to many," he said, "all that we know about it is that we know nothing. . . . Most biologists . . . having agreed that life's advent was at once the most improbable and the most significant event in the history of the universe, are content for the present to leave the matter there."

Hopkins's assessment was not entirely true. The subject had been pursued intently by a handful of scientists, notably the Mexican Alfonso Herrera and the Frenchman Stéphane Leduc. And in the 1920s, a pair of scientists had independently come up with remarkably similar and groundbreaking theories about the origin of life whose impact would be felt throughout the rest of the century and beyond. Neither theory initially attracted much attention. But in the decades to follow, they would form the basis of a renewed scientific search for answers.

Like Darwin and Wallace, both men had come up with their theories independently, each unaware of the work of the other. One was a young scientist from the Soviet Union whose work was virtually unknown in the West. The other was an eccentric whose skill as an essayist had earned him a reputation as England's most popular writer of science. But they had a couple of things in common. In their theories, both men made use of the vast advances in understanding the ancient Earth and ancient life that had been achieved by the 1920s. And both were impassioned Marxists for whom politics was almost as important as science.

Eventually, their ideas came to be merged into a single theory that represented a monumental reconfiguration of the understanding of how life appeared on the planet Earth. It would come to be called the Oparin-Haldane hypothesis, named for the Russian Alexander Oparin and the Scotsman J. B. S. Haldane.

JOHN BURTON SANDERSON HALDANE—Jack to his friends and J. B. S. to everyone else—was a contrarian who enjoyed setting himself apart from the crowd. In his later years, when his theories on the origin of life gained acceptance in mainstream science, he joked that it made him wonder what he'd gotten wrong. He had little time for social propriety. He liked to shock people, brazenly boasting about sexual conquests that those who knew him best seldom believed actually happened. Though he was a handsome man—strikingly so during his younger years—Haldane was always insecure around women, probably as a result of a childhood of bullying brought on by the intellect that distanced him from children his own age. He reveled in his peculiarities, and exaggerated them, but they were quite real.

Haldane was always a bit of a pyromaniac, notorious in his army years for carrying explosives and loose matches together in his pockets. "Those who lived in close contact with him regarded his presence in the mess or officers' quarters with some degree of suspicion or fear," one of his military colleagues later recalled. Haldane never lost his love of explosions and fires, something he had in common with English folk legend Guy Fawkes, whose birthday he shared. When lighting the tobacco pipe he habitually smoked, Haldane would let the match burn down until the tips of his fingers were black. He was one of those individuals who had what Thomas Huxley would have called "a screw missing." Haldane compensated for this deficit by having a host of screws other people simply did not possess.

Haldane's remarkable intelligence had been apparent from a very young age. After injuring his forehead when he was four, he asked a doctor whether his blood was "oxyhemoglobin or carboxyhemoglobin." Such terminology was common in the household in which he was raised. His father, Lord Haldane, was a Scottish aristocrat and noted Oxford physiologist who had constructed a private laboratory in their home. Whenever possible, Lord Haldane tried to involve his children in his scientific work, even if it sometimes meant putting them in danger. He often used them as human guinea pigs to test the effects of gases. Many years later, when investigating the effects of different gases on submariners for the British government during World War II, J. B. S. Haldane would continue his father's practice by using himself as a test subject.

Much of Lord Haldane's work revolved around the study of gases encountered by British coal miners, work that would one day lead to his design of Britain's first effective gas mask during World War I. During Lord Haldane's investigative trips into some of Britain's most dangerous mines, he sometimes brought his son, who would be lowered by bucket into the miserable crawlspaces the miners used. On one such trip, father and son were nearly killed by a pocket of deadly combustible methane gas. They were saved

J. B. S. Haldane in 1941.

by Humphry Davy's safety lamp, which alerted them to the presence of the gas when it reacted by shutting itself off. The trips were J. B. S. Haldane's first exposure to the dangers and miseries heaped upon the British working classes, on whose behalf he would agitate for the rest of his life.

During his college years at Eton, Haldane so distinguished himself in the sciences that many of his fellow students felt as if he was teaching the teachers. Yet halfway through completion of an undergraduate degree, he changed his course of study to the classics. This lack of a formal specialization had a lot to do with the way Haldane approached science. Since the time of the scientific revolution in the sixteenth century, science had slowly but steadily become increasingly specialized, being studied and taught in separate faculties, with physics, chemistry, medicine, and other fields becoming more and more compartmentalized. By the twentieth century, the barriers between these disciplines were firmly in place, but Haldane could hold his own on nearly any scientific subject.

With the outbreak of World War I, Haldane joined the Scottish Black Watch regiment. Commissioned as a lieutenant, he found himself in command of one of the first trench mortar detachments. These early experimental mortars were little more than explosive-laden stovepipes used to harass the enemy through constant bombardment. They were exceedingly unpredictable, "almost as dangerous to their users as the enemy" in the words of the official regimental history. Haldane thrived on such risks, actually encouraging his men to smoke around their mortars. He later recalled that he "thought it important that we should have absolute confidence in ourselves and in our weapons."

In combat, Haldane distinguished himself with a bravery that sometimes bordered on insanity. At night, during lulls in the fighting, he would sneak out into the no-man's-land between the opposing armies' trenches hoping to catch snippets of the enemy's conversations. In letters home, he compared the thrill of war favorably to the feelings of exhilaration he had experienced in the mines that he explored with his father, when a single lapse could lead to the worst kind of catastrophe. Sir Douglas Haig, com mander of all British expeditionary forces fighting in France, called Haldane "the bravest and dirtiest officer in my Army."

After he was wounded in an artillery barrage, Haldane was sent to Iraq to fight Germany's Turkish allies. When he returned to Britain at the war's end in 1919, he carried few of the deep psychological scars that haunted millions of his fellow soldiers. Haldane would tell people that he had enjoyed the war, "loved it," in fact. He chalked this fact up to evolution, his ancestry from early humans for whom the urge to kill would have been an evolutionary advantage. Eventually, after witnessing the horrors of war as an observer rather than a participant, he became uneasy about his own, favorable experience of World War I. In his final years, he came to embrace Ghandian nonviolence.

After the war, even though he never took any kind of scientific degree, Haldane was offered a teaching post at Oxford in any subject he desired. He chose physiology. Later, he accepted an appointment at Trinity College, Cambridge, where he met and married a young Jewish reporter for the *Daily Express* named Charlotte Burghes, a socialist and a feminist. In 1926, their first year together, they traveled to the Soviet Union at the invitation of the Russian botanist Nikolai Vavilov, the president of the Soviet Academy of Agricultural Sciences. Haldane found himself drawn to the young Soviet state, and in the years to follow, he drifted slowly toward communism, particularly after the spread of European fascism and the outbreak of the Spanish Civil War.

When fighting broke out in Spain in 1936, Haldane determined to help the Republican side. Decked out in a visored motorcycle cap, breeches, and a black leather jacket, he flew to Madrid, where he hoped to advise the government on dealing with air raids and the possibility that Francisco Franco's fascist rebels would resort to chemical warfare. At one point, his fanciful dress got him arrested by Republicans who mistook him for a fascist spy.

While in Spain, Haldane began sending back accounts of his experiences for publication in the communist *Daily Worker*. When he returned to Great Britain, he became the paper's science correspondent and, eventually, the chairman of its editorial board. His involvement amounted to a huge coup for the Communist Party. Haldane had already acquired a reputation as a superb essayist. He had a unique ability to present complex subjects in a simple way that the average person could read and comprehend. Haldane's admirer Arthur C. Clarke, the author of *2001: A Space Odyssey*, called him "perhaps the most brilliant scientific popularizer of his generation."

Haldane counted many science fiction writers among his friends. One of the best known was Aldous Huxley. The influence of Haldane's theories could be found in Huxley's *Brave New World*, and it was Haldane who first used the term "clone." Many of Haldane's topics seemed so far out—the concept of in vitro fertilization and test-tube babies, the idea that power could be harnessed from hydrogen—that his early writings were sometimes miscast as science fiction. And because of his popularity among lay audiences, scientists found it easy to underestimate his impact as a serious scientist and researcher, particularly in the field of evolutionary theory and the burgeoning field of genetics.

Haldane's most lasting impact was on the subject of the origin of life. His first and most important essay on that topic, entitled, simply, "The Origin of Life," was written in 1929. He began the essay with a defense of his materialistic vision of the world, lampooning the widely held notion that Pasteur had somehow proved that life could not spring from nonlife. Such a notion could be held only by those who "clung to the supernatural." Either an act of abiogenesis had at some point occurred or "a living thing is a piece of dead matter plus a soul (a view which finds little support in modern biology)." Perhaps the most striking feature of Haldane's theory was that it conjured up a well-supported vision of what the Earth would have looked like when life first appeared billions of years in the past. It was a world altogether different from the one Charles Darwin had once speculated about. Where Darwin had imagined a "warm little pond," Haldane imagined a "hot dilute soup"* that would have been hostile to any form of life we would be familiar with.

---

* For an exhibit on the origin of life in the 1980s, the Smithsonian National Air and Space Museum played on Haldane's theory by enlisting Julia Child for a short film in which the famous television chef taught viewers how to make "primordial soup."

Haldane's model for the origin of life reflected new advances in geology, biochemistry, and our understanding of the evolutionary process. It was a modern theory that could be researched and built upon, far more sophisticated than the simple concepts of spontaneous generation that had dominated earlier thinking. But it took a while for the implications of his essay to sink in among serious scientists. At first, it was often brushed aside as mere speculation. Gradually, though, the plausibility of Haldane's theories began to gain traction. They garnered even more credibility as scientists in the West became aware of a similar but more elaborate hypothesis that had been worked out in the Soviet Union.

Though few scientists in the West could understand Russian, news started to spread of a short book on the origin of life that had been published in 1924, five years before Haldane's essay. Its author, Alexander Oparin, had worked out a model remarkably similar to that of Haldane. And Oparin's treatment of the question had been more exhaustive and scientifically rigorous. Unlike Haldane, who had moved on to other subjects, Oparin spent the rest of his life honing and developing his theory, and by 1936, when an English translation of Oparin's second, longer book on the subject began to arrive in Western universities, students of evolutionary biology rushed to get their hands on copies of what was to become the seminal work on the subject for much of the century. On the seventy-fifth anniversary of its publication, *Nature* magazine remembered Oparin as deserving of a place in "the pantheon of the twentieth century's greatest scientists," though few in the West, outside of the narrow disciplines touched by his work, would recognize his name.

ALEXANDER OPARIN was the child of a Russia that had almost entirely disappeared by the time he left his mark on science. He was born in a small town along the Volga River north of Moscow called Uglich. It was a rural community that might have been lifted from a nineteenth-century Russian novel, filled with wooden houses and horse-drawn carts carrying crops to market. Oparin's simple background often surprised those who knew him later in life. He had a fondness for well-tailored suits, always accompanied by a trademark bow tie that seemed outlandish in postrevolutionary Russia. His cosmopolitan attire and his peculiar Yaroslavl accent, with its pronounced emphasis on the "O" sound, often led his countrymen to mistake him for a foreigner.

Marxism and a faith in evolutionary natural history had come early to Oparin. Both were introduced to him by his boyhood idol, the Russian botanist Kliment Timiryazev. Oparin had picked up an interest in botany during his childhood in Uglich, where he often spent his days roaming the countryside in search of new plants for his collection. He read Timiryazev's book, *The Life of the Plant*, so many times that he felt as if he could almost recite it by heart. During the waning years of tsarist Russia, when Oparin was a teenager living in Moscow, he would often visit the Moscow Polytechnic Museum, where Timiryazev would regularly lecture in a splendorous amphitheater designed by the talented Ippolit Monighetti, the favorite architect of the Romanovs.

Timiryazev had by then become known as one of the earliest Russian converts to Darwinian evolutionism. As a young man, he had even made a pilgrimage to Darwin's house in Downe not long after the publication of *Origin*. Told that Darwin was ill and not receiving visitors, Timiryazev had rented a room above the local pub. For a week, he returned every day to sit on Darwin's stoop until his hero finally granted him a pleasant afternoon of walking and conversation. Back home, Timiryazev became the country's most prominent advocate of Darwinian evolution, occupying a role in Russia similar to Huxley's in Great Britain.

For Timiryazev, Darwinian evolution was more than just a scientific theory. It was a revolutionary force, materialistic and atheistic, with implications that could be carried into social and political spheres. Such connections usually found their way into Timiryazev's spellbinding science lectures, and eventually cost him his position at Russia's most prestigious center of higher learning, Moscow State University. Listening to Timiryazev at the Polytechnic Museum turned Oparin into a devotee of the elder scientist's radical politics and evolutionary leanings.

Oparin was captivated by Timiryazev's stories about the great Charles Darwin. But there was one thing that nagged at him from the very first moment he heard Timiryazev explain Darwin's concept of evolution. Darwin had simply skipped over what Oparin saw as the most important part of the materialistic theory of evolution: the origin of life. Later, when he became a professor, Oparin would tell his students that "Darwin had written the book, but it was missing the very first chapter." This was a void Oparin would spend most of his life trying to fill.

OPARIN WAS BARELY TWENTY when World War I broke out, and he spent most of the war studying plant biology at Moscow State University. After graduating in 1918, he started a mentorship under the eminent plant biologist Aleksei Nikolaevich Bakh, who had become a legend among Russia's socialist revolutionaries. As an old-guard member of the leftist People's Will group that had assassinated Alexander II, Bakh authored the most iconic piece of propaganda produced during the Russian Revolution, *Tsar-Golod* ("Tsar of Hunger"), an eloquent denunciation of the Romanov dynasty and capitalism.

Oparin finally met Bakh in Moscow, shortly after Bakh's return from exile in Switzerland and about a year after the abdication of the last absolute monarch in Europe, Tsar Nicholas II. Vladimir Lenin had seized power in a coup that would be remembered as the Great October Socialist Revolution, and the streets of Moscow were abuzz with revolutionary workers and menacing groups of militia calling themselves Red Guards. Under the Bolsheviks, Bakh found himself gradually elevated into an important leadership role in postrevolutionary Russian science. He and Oparin founded the country's Russian Academy of Sciences' biochemistry institute. Bakh was named the director, followed by Oparin after his mentor's death.

Much of Oparin's early work revolved around food production, a huge priority in the chaotic first years of the regime and something Oparin threw himself into wholeheartedly. But Oparin never stopped working to find answers to that seminal question that had so intrigued him since he had attended Timiryazev's lectures: the origin of life. He wrote his first work on the subject in 1919, but it was rejected by state censors. In the years immediately following the October Revolution, much of the tsarist state apparatus was initially left intact, including the censorship boards, which were still deeply sensitive to anything that might contradict the official line of the Russian Orthodox Church. Oparin would one day see that rejection as a boon. It enabled him to sharpen a more sophisticated argument and theory.

By 1922, Oparin was presenting those new ideas on the subject of the origin of life to Soviet scientific bodies. In 1924, he sat down to write a book that would lay out what had by then blossomed into a grand theory. Like Haldane, Oparin tackled the problem in a way that was fundamentally different from the approach his scientific predecessors had taken. Scientists like Thomas Huxley or Henry Bastian had worked on the assumption that the first life appeared on an Earth not so different from the Earth as it still was. And this process, they assumed, had happened rather quickly.

Oparin and Haldane, on the other hand, were searching for answers about something that had happened, they believed, at least many hundreds of millions of years in the past on a planet that could scarcely be recognized and under conditions they could only creatively imagine. But both men had a great deal of evidence that simply had not been available to earlier evolutionists. Even though the study of the origin of life had stagnated for the previous four decades, the understanding of the conditions under which the first life appeared had changed dramatically. For the first time, scientists were beginning to appreciate that the Earth was much older than any of their predecessors had imagined, and that life had existed on the planet throughout most of its history.

TIME HAD ALWAYS BEEN an enigma for Charles Darwin. He believed that the pace of evolution based on natural selection was extraordinarily slow, with species being transformed inch by inch through countless generations full of evolutionary dead ends and long periods of stagnation. It was hard to account for the evolution of the simplest microorganisms into complex species like human beings in a time frame that people could accept.

The problem persisted, even though Buffon's once-radical guess at the Earth's age seemed absurdly timid by Darwin's era. In the first edition of *On the Origin of Species*, Darwin had given his own reckoning of the Earth's age. Like the estimates arrived at by Buffon and Ussher, Darwin's figure was exact almost to the point of being silly. He estimated that the Earth was 306,662,400 years old, a figure based on his assessment, from geological clues, of the age of southern England.

Darwin's estimate drew the scrutiny of the Irish physicist William Thompson, usually remembered by the title he acquired later in life, Lord Kelvin. Kelvin was one of the most accomplished and publicly revered scientists of the age. His role in building the first transatlantic telegraph had brought him enormous fame, fantastic wealth, and ennoblement. He had also helped formulate the first and second laws of thermodynamics, which he had used to produce his own estimate of the age of the Earth. Like Buffon's estimate, Kelvin's was based on how long it would have taken the Earth to cool to its present temperature. Because he was unaware of the process of radioactive decay, which accounts for much of the heat generated below the Earth's surface, Kelvin assumed that the Earth was a rigid sphere that had been cooling since its inception, and that he could judge its age by comparing the Earth's exterior temperature to those taken from its interior.

Three years after the publication of *Origin*, Kelvin postulated that the Earth was between twenty million and four hundred million years old, but he revised the estimate downward in the coming years, largely to correspond with his much lower—and now recognized as vastly wrong—estimates of the age of the sun. By 1897, he had settled on twenty to forty million years—"much nearer 20 than 40." Thomas Huxley attacked Kelvin's methods as faulty, but even Darwin's own son, the astronomer George Darwin, had put forth a relatively low estimate of fifty-six million years, which he based on his calculation of how long it would have taken the Earth to settle into its current twenty-four-hour cycle of daily rotation. The Earth's age remained an important—and strongly contested—subject of debate through the end of the nineteenth century.

In the face of so much dispute about the true figure, Charles Darwin removed the reference to the Earth's age from the second and all future editions of *Origin*. The question puzzled him throughout his life and hindered acceptance of the slow process of natural selection as the principal agent of evolution. Even Darwin's staunchest supporters conceded that natural selection would likely have required hundreds of millions of years, but such a time frame was difficult to reconcile with the best estimates of how long evolution would have had.

Then, a discovery in France set in motion a chain of events that upset all the assumptions about the Earth's age. In 1896, a year before Kelvin gave his final estimate, a French physicist named Henri Becquerel happened to leave a packet of uranium salts on a Lumière photographic negative. Later, Becquerel returned to find an image of the packet burned onto the negative as if it had been photographed. By placing objects between his uranium and the negative, he found he could produce images of anything. The only conclusion he could draw was that invisible rays of energy were being emitted from the rock. Three years later, Marie Curie discovered the elements polonium and radium. She coined the term "radioactivity" to describe the mysterious energy they emitted.

In a remarkably short period of time after the discovery of radioactivity, physicists developed methods to measure the age of rocks based on the decay of radioactive elements. Every rock is made up of chemical elements, some of which are present as a mixture of isotopes, atoms of the same element that have different numbers of neutrons in their nuclei. Some isotopes are unstable and radioactive, and they are constantly, albeit slowly, decomposing into new, lighter elements. The length of time it takes for half of those molecules to decay into a new element is called half-life. Though each rock is initially endowed with some ratio of the isotopes of its various component elements, over time some of these undergo radioactive decay to give new ratios. By measuring these ratios, geologists learned to calculate how much time had passed since the rock had formed. The process came to be called "radiometric dating."

In 1907, the chemist Bertram Boltwood published the results of a radiometric study of twenty-six rocks, one of which he found to be a staggering 570 million years old. As radiometric techniques were refined, the age of Boltwood's oldest rock increased to 1.3 billion years. Other geologists were finding rocks even more ancient, including one from Ceylon that was 1.6 billion years old. It took until the middle of the twentieth century for most scientists to agree on a figure for the age of the Earth of about 4.5 billion years. Even by the time of Oparin's return to Moscow, most scientists understood that the Earth was vastly older than anyone a century earlier would have imagined it to be.

Yet the question still remained of just how long *life* had existed on the Earth. Huxley had postulated that abiogenesis was extremely rare, something that may have happened only once and only by a confluence of conditions and chance that made it extremely unlikely to happen again. It was possible that Earth had been lifeless for most of its existence. This was, in fact, exactly what the incomplete fossil record seemed to suggest.

During the first half of the nineteenth century, geologists had to make do with fossils found in irregular settings caused by the coming together of ideal geological circumstances, and then only those exposed at the surface were easily examined, such as the fossils Darwin had found on the volcanic island of St. Jago. The industrial revolution began to change all that. As long canals were constructed throughout the British Isles, connecting ports and coal-mining regions to inland industrial centers, geologists were left with deep, clean gashes into the Earth exposing strata that had accumulated over eons. They began to appreciate that certain fossils were always found in certain stratigraphic layers, and never in others. They didn't yet know how old those layers were—such knowledge would come only with radiometric dating—but they did understand that certain layers were older than others.

Eventually, they divided the time represented by the different layers into two long eons. The shorter and more recent was called the Phanerozoic eon, Greek for the "age of visible life." The Phanerozoic was subdivided into even shorter geological periods, the earliest of which was called the Cambrian, a name coined by Adam Sedgwick after "Cambria," the Latin name for Wales, where many of the first samples from that period were found. The Phanerozoic was, in turn, preceded by the older and vastly longer—and less imaginatively named—*Pre*cambrian eon.

When Darwin wrote *Origin*, all of the fossils scientists had acquired came from the shorter, more recent Phanerozoic eon, which we now know comprises a mere 15 percent of the history of the Earth. He addressed his dilemma in *Origin*: "If the theory [evolution] be true, it is indisput-able that before the lowest Cambrian stratum was deposited, long periods elapsed . . . and that during these vast periods, the world swarmed with living creatures . . . why we do not find rich fossiliferous deposits belong-ing to these assumed earliest periods prior to the Cambrian system, I can give no satisfactory answer. The case at present must remain inexplica-ble." The answer would eventually be found nearly a half century later and half a globe away, in the United States, where a young geologist named Charles Doolittle Walcott was fast becoming the most important fossil hunter in the world.

Raised in Rhode Island by a single mother, Walcott never finished high school and never spent a day in college. As a teenager, he had become a professional fossil collector, selling his finds to universities akin to the way Alfred Russel Wallace had once supported himself by collecting live spec-imens. At age twenty-six, Walcott was hired as the assistant to the chief geologist of the state of New York, James Hall, a man as famous for his tyrannical disposition as his expertise in paleontology—and he was very famous for his paleontology.

Hall let Walcott in on one of his most intriguing discoveries: a strange-looking reef in a riverbed near the town of Saratoga, decorated by round patterns imprinted on limestone, each about a meter wide. Hall was convinced that the shapes were biological in origin and that they had been left by colonies of millions of microscopic algae. He called his hypothe-sized microbes *Cryptozoon*, "hidden life." The trouble was that he didn't have an actual fossil. Even in the twenty-first century, identifying a micro-scopic fossil is a painstakingly difficult task. Microbial cells are not that different in shape and size from all manner of natural nonliving particles. Since they have no skeleton, they do not fossilize well. Debate often hangs on contextual clues such as the type of environment where the surround-ing rock was deposited and the ratios of various isotopes in elements such as carbon and sulfur, which can hint at the hand of biology. While mod-ern micropaleontologists have sophisticated equipment for determining whether fossils are biological in origin, for Walcott and his contemporaries these techniques did not yet exist. There was little scientific acceptance of the evidence for *Cryptozoon* being of living origin. Walcott needed better microscopic evidence.

Three years later, with Hall's recommendation, Walcott was appointed to the newly formed US Geological Survey (USGS). Soon he was heading west to explore one of the greatest natural wonders of North America, the Grand Canyon, about which remarkably little was then known.

Charles Doolittle Walcott at the Grand Canyon.

The Grand Canyon turned out to be a paleontologist's dream. For seventeen million years, the Colorado River had chiseled a course deep into the hard, rocky ground, leaving a majestic gash 277 miles long and over a mile deep. Though it is second in size to Nepal's Kali Gandaki Gorge, the Grand Canyon's bareness made it unrivaled in potential for study by a paleontologist like Walcott. Its features were not hidden by rich vegetation as they were in the Kali Gandaki, or even the smaller foothills of England and Scotland that had been explored by most of the century's greatest fossil hunters. Its walls resembled the clean, layered faces of a canal, except it was a canal that cut a mile beneath the surface and through two billion years of rock formations.

The leader of the USGS expedition, John Wesley Powell, realized the Grand Canyon's potential as an archaeological site. He put Walcott to work doing what he did best, hunting for fossils. Walcott soon found signs of life similar to Hall's *Cryptozoon*. More significant, he found them in what were almost certainly Precambrian rocks. In 1891, Walcott wrote that "there can be little, if any, doubt" that life indeed existed in Precambrian seas, but not until 1899 did he find the definitive evidence he was looking for. Twenty years after first being intrigued by Hall's *Cryptozoon*, Walcott discovered in the Grand Canyon the fossilized remains of microscopic, single-celled algae that he named *Chuaria* after the rock strata in which they were found. Though his discovery remained controversial into the twentieth century, *Chuaria*-like fossils have now been dated to as far back as 1.6 billion years. Walcott had finally found the answer to Darwin's missing piece of the fossil record. Eventually, even older fossils would be found, and scientists would come to accept that simple life-forms had existed for at least 3.5 billion years of the Earth's 4.5-billion-year history.

BY THE TIME Oparin and Haldane set about formulating their theories of the origin of life, they understood that the Earth was vastly older than any of their predecessors could have imagined. This was a crucial point because it meant that the environment of the planet when life first arose was probably nothing like that of the modern world. Oparin and Haldane could throw out the old assumptions about spontaneous generation—that any appearance of life from nonlife should be repeatable in an environment that would now be familiar to us—and instead speculate on what kind of world it would have taken to produce life.

As Oparin set about working out his theories in the 1920s, and particularly in his 1936 book, he could draw on those facts to paint a new picture of a young Earth as it existed hundreds of millions or even billions of years earlier. It was an Earth so vastly different that it might as well have been an alien planet, its atmosphere in particular would have been different.

Figuring out which elements had to be present was the easy part. Broken down to their most basic chemical parts, all living things are remarkably similar. They are also remarkably simple. From the smallest bacteria to the cells of the most complex species, living organisms are made primarily of carbon, hydrogen, oxygen, and nitrogen, the four basic elements of life that chemists often refer to by the acronym CHON. Other elements are found in trace amounts, most important among them sulfur and phosphorus, but about 98 percent of every living thing is made up by weight of the four elements C, H, O, and N.* Each of those elements was almost certainly abundant on Earth, and just about everywhere else. They are, in fact, four of the seven most common elements found in the universe.

The hard part was understanding how these elements combined to form the more complex molecules required for life. The CHON elements may have been present, but their form on the primitive Earth was still an open question. Was the oxygen present only in water (H_2O), or was it also free as O_2 gas in the atmosphere, as is true of the modern Earth? To understand how life may have come about required first understanding what kinds of chemical *compounds* were available at the time.

Oparin began with an assumption that there was no free oxygen gas in the primitive atmosphere. From astronomical observations that had been made of Jupiter, Oparin deduced that the early Earth's atmosphere

---

* Various other acronyms are sometimes used, reflecting the other elements. SCHNOPs and SPONCH highlight the presence of sulfur and phosphorus.

was filled with methane and ammonia. It was also an environment bathed in external energy, bombarded from above by cosmic rays and ultraviolet radiation, unchecked by the modern Earth's ozone layer. The surface was wracked by constant volcanic activity far beyond anything experienced today. Excited by the bombardment of solar radiation and heated by the energy released by volcanoes, the atmospheric gases would have broken down into their constituent parts. These would have recombined into new compounds, some of which would have dissolved into the vast seas that covered most of the planet. This long chain of chemical events would have led to the synthesis of organic compounds and, eventually, to some sort of precellular structure that represented an intermediate stage between non-life and life. Haldane's vision of the early Earth was strikingly similar, and the areas of consensus between the two men's theories formed the basis of the Oparin-Haldane hypothesis.

In the decades to follow, as scientists learned more about the geological and astronomical conditions that had existed when life first appeared on Earth, several elements of the Oparin-Haldane hypothesis proved remarkably resilient. Geochemistry, the study of the ways the fundamental laws of chemistry can be used to explain planetary processes, would prove by the end of the century that the early Earth's atmosphere did not, in fact, contain much oxygen—a condition that lasted for almost two billion years after the Earth's formation, until biology invented oxygen-generating photosynthesis. And the lack of oxygen meant that the atmosphere would have had little ozone, leaving the Earth unprotected from ultraviolet radiation from the sun.

This last fact, the high flux of energy, was extremely important in both Haldane's and Oparin's theories. It was the driving force for the natural synthesis of organic compounds. The compounds intermingled, forming simple molecular aggregates. These were simpler than any single-celled organism we would know of today, but complex enough to convert organic compounds into more copies of themselves. Some of them attained enough complexity that Haldane called them "half-living." Oparin called these molecular aggregates "coacervates."

At this point, Haldane's and Oparin's visions diverged in ways that would be increasingly significant in the decades ahead. Each man had a different idea of what made something living. Oparin saw the key as cellular metabolism, the collection of chemical reactions that transform external foodstuff into living material. Life for him was a chemical process, and its essential components were proteins that helped these processes occur. This school of thought came to be known as the "metabolism first" tradition.

For Haldane, the key to life lay in the gene. His concept of an intermediate stage between life and nonlife was influenced by the phenomenon of viruses, which scientists then understood in only a rudimentary fashion.* Considerably smaller than bacteria, viruses would not be seen under microscopes until 1933, and there was considerable disagreement as to whether or not viruses were, in fact, living. Haldane was particularly intrigued by something called a bacteriophage, a virus that infected bacteria and seemed to Haldane to hold characteristics that might lead it to be called "half-living." In 1915, the French-Canadian microbiologist Félix d'Herelle was struggling to understand why water from India's Ganges and Yamuna Rivers seemed to have the remarkable trait of being able to protect people from cholera. Both rivers were filthy with sewage and teemed with harmful bacteria. D'Herelle found that the rivers also contained a remarkable "bacterium eater," which some observers soon claimed possessed the ability to self-replicate within cells.

THE UNIQUE PERSPECTIVES on biology held by Haldane and Oparin were shaped by the very different scientific worlds in which the two men worked. In the Soviet Union, the field of genetics was increasingly seen as a "bourgeois science," based on a kind of "survival of the fittest" model that was anathema to the Marxist ideal. In the West, the study of genetics was rev-

---

* Although Pasteur had developed antiviral vaccines, he was unaware of the existence of actual viruses. The first experimental evidence for viruses was not available until 1892, when Russian botanist Dmitri Ivanovsky showed that sap from infected tobacco plants that had been put through a filter capable of removing all the bacteria still had the power to spread disease.

olutionizing biology, and Haldane was one of the field's most important elaborators and theorists.

Though the basic tenets of genetics had been elaborated in the mid-nineteenth century by the Franciscan friar Gregor Mendel, his work had gone almost entirely unnoticed until 1900, when several scientists independently realized the importance of what Mendel had discovered. The concept was quickly embraced, and by 1906, scientists were using the term "gene" to describe the unit of inheritance.

Haldane had been on the forefront of genetics since the very beginning. In 1901, when he was nine, his father had taken him to one of the earliest lectures on Mendel's theory. Genetics became one of his favorite subjects and remained so for the rest of his life, occupying much of his writing and experimental work. Haldane was instrumental in establishing the interlocking relationship between Darwin's theory of natural selection and Mendel's theory of heredity.

Acceptance of genetics in the Western world was swift. The basic concepts were easy enough to explain and to prove, even to a child of nine like Haldane. Yet, by the mid-1920s, the study of the gene was increasingly being brushed aside in the Soviet Union, where Joseph Stalin was forcing the scientific establishment more and more to conform to his totalitarian view of Marxism, in which genetics and "survival of the fittest" were seen at odds with the utopian vision of absolute equality.

The worst excesses of the period were personified in one scientist in particular, the agronomist Trofim Lysenko, who for two decades almost single-handedly managed to retard the science of genetics in the Soviet Union. The son of an illiterate peasant farmer, Lysenko was a little-known scientist working in the Ukraine when *Pravda* first profiled him in a 1927 article on Soviet farming titled "The Fields in Winter": "If one is to judge a man by first impression, Lysenko gives one the feeling of a toothache. All one remembers is his sullen look creeping along the earth as if, at the very least, he were ready to do someone in." As Lysenko rose up the ranks of Soviet bureaucracy through brutal Machiavellianism and Stalinist sycophancy, the description proved to be remarkably prescient.

Lysenko was one of the holdouts of Lamarck's increasingly marginalized theory of acquired characteristics. Lamarckian theory was central to Lysenko's attempts to breed crops that would be more resilient in winter. His claims that he had developed a method to convert spring wheat into winter wheat proved disastrous for Russian agriculture, which was already reeling from the inefficiencies of mass collectivization. In the years to follow, millions died of starvation.

Yet Lysenko shrewdly used his peasant background to endear himself to Stalin and quickly rose to a commanding position in the Soviet biological sciences. The results of his stewardship were catastrophic. Most scientists working under Lysenko caved in to the climate of fear, adjusting their theories to conform to Marxist dogma as expounded by men like Lysenko. Those who did not sometimes paid with their lives.

One of Lysenko's victims was Nikolai Vavilov, the geneticist who had sponsored Haldane on his first trip to Russia. Vavilov was one of the most prominent Mendelian holdouts to resist Lysenko, though he otherwise bent over backward to win the Ukrainian's favor. Vavilov had even sponsored Lysenko's membership in the Ukrainian Academy of Sciences. In 1940, Vavilov was nevertheless arrested on charges of espionage. A sentence of death by firing squad was commuted to twenty years in prison. He died two years later of starvation. By 1948, nearly all geneticists were being called before Communist Party meetings and forced to recant.

Oparin's role in Lysenko's reign of terror remains a dark stain on his career. The two men shared summer dachas together and were known to be friendly, though in the Orwellian world of Stalinist Russia, it was hard to know how much of their friendliness was genuine. Marxism certainly affected Oparin's work. Especially in his early years, Oparin really did seem to be the kind of true-believer communist scientist that became so rare in the later days of the Soviet Union, after years of oppression had squeezed out what little authentic idealism the scientific community could muster. In some ways, Oparin's broad approach to the question of the origin of life could be traced to his embrace of Marx's historical dialectic. He was probably helped by the fact that scientists in the Soviet Union tended to be less trapped into a single discipline than their Western counterparts were. In his emphasis on the slow but deliberate evolutionary steps that led from nonlife to life, Oparin tried to keep close to a theory on the origin of life that had been posited by Marx's collaborator Friedrich Engels in his book *Dialectics of Nature.* Though Oparin never incorporated Lysenko's theories into his own work, until the 1950s he shied away from touching on the questions of genetics that would come to dominate origin-of-life research by the end of the century.*

Many scientific dissidents saw Oparin as, at worst, a close ally of Lysenko and, at best, an enabler. His many friends in the increasingly internationalized community of scientists searching for the origin of life found his complicity in the darkest excesses of Lysenkoism hard to reconcile with the kind and warmhearted man who managed to charm almost everyone he encountered.

FOR HALDANE, the long descent of the world's most prominent communist state into the madness of Lysenkoism proved too much to bear. In 1948, in the wake of the Lysenko-inspired war on genetics in the Soviet Union, Haldane had been forced to give a private recantation of genetics at a British Communist Party meeting. He quit the party the following year.

Yet Haldane never abandoned the general framework of his left-wing ideals. Disgusted with the British invasion of Egypt during the Suez Crisis in 1956, Haldane emigrated to India, where he found the moderate socialism of the Indian Congress Party to his liking. Haldane took to his adopted country, becoming a vegetarian and dressing himself in a traditional skirt-like dhoti, which occasionally caused him to be mistaken for a guru. He liked to joke that his emigration was spurred by the desire to stop wearing socks. "Sixty years in socks is enough," he once wrote. After his death, a prominent road in front of South Asia's largest science center—Science City, near Kolkata—was named after him.

In 1963, Haldane was asked to speak at a series of conferences and meetings on the subject of the origin of life that were being held in the

---

* In a 1971 interview with science historian Loren Graham, Oparin defended his actions during the Lysenko period as a practical necessity. "If you had been there during those years," he said, "would you have had the courage to speak out and be imprisoned in Siberia?"

United States. One of the events was at the Institute of Biological Sciences in North Carolina. Since state law prohibited members of the Communist Party from speaking at functions that had received state funding, Haldane was asked whether he was now or ever had been a member of the Communist Party. He refused to answer the question, pointing out in his reply that the Soviet Union did not ask visiting lectures if they were members of the Conservative Party. He went on to say that he would "use the incident for propaganda against the present set-up in your country, which is of course, in flat contradiction of the principles laid down by your Founding Fathers."

Despite the episode in North Carolina, Haldane managed to complete the rest of his journey without incident, including his final stop, a conference on the origin of life in Wakulla Springs, Florida, a rural town outside of Tallahassee. The conference organizer, a NASA-supported chemist named Sidney Fox, had also invited Oparin. Although they had corresponded for decades, Oparin and Haldane had never actually met in person.

Oparin was the opening speaker, and Haldane introduced him. "I suppose that Oparin and I may be regarded as ancient monuments in this branch of science, but there is considerable difference," Haldane said. "Whereas I know nothing serious about it, Dr. Oparin has devoted his life to this subject." Haldane left the conference early, after experiencing abdominal pains. At a hospital in Tallahassee, he was told he had cancer. He died within the year.

THE HYPOTHESIS developed by Oparin and Haldane created a new framework for understanding the origin of life. It was a modern attempt to imagine the environmental conditions in which life might have arisen, conditions extremely different from those imagined by most of their predecessors. Although a full elaboration of the true workings of the cell would wait until the second half of the century, both men drew upon an increasingly sophisticated understanding of cellular mechanics. Still, despite proposing ideas that represented a major theoretical breakthrough, they did little to advance the subject experimentally. Haldane never engaged in any experiments on the subject, while Oparin's attempts yielded little of consequence and were not generally held in wide regard. But their ideas galvanized a new generation of researchers into the field. Soon their hypothesis was buoyed by one of the most famous experiments of the twentieth century.

A LABORATORY EARTH

If there were in nature a progressive force, an eternal urge, chemistry would find it. But it is not there.

—WILLIAM JENNINGS BRYAN, 1925,
undelivered closing argument at the Scopes Monkey Trial

I N THE SPRING OF 1953, a young graduate student by the name of Stanley Miller walked into the Kent Hall lecture room at the University of Chicago. Miller was nervous. He was just twenty-three and was about to address some of the most accomplished scientists in the United States. Chicago's chemistry department was one of the most prestigious in the world. During World War II, the university had been a focal point of the American atomic weapons program. Scores of top scientists had stayed on or had become affiliated with the university after the war.

Many of the most important figures in the Manhattan Project were in the audience for Miller's lecture, including several winners of the Nobel Prize and several more who would win that highest honor of science in the years to come. Most prominent among them was the man sometimes called the "father of the nuclear bomb," Enrico Fermi. In 1942, Fermi had built the world's first nuclear reactor just down the road, under the bleachers of the university's old football stadium, Stagg Field. It was a remarkably simple device, little more than uranium pellets and graphite blocks arranged in a pile, from which it took its name, Chicago Pile-1.

Over the previous year, Miller had been working on the question of the origin of life as part of his PhD thesis project. Since the fall of 1952, word had spread throughout the university that Miller had successfully executed an experiment showing how life might have arisen under the primitive conditions of the early Earth. Using nothing more than a glass apparatus, an electric Tesla coil, and some simple gases, he had created amino acids, the basic building blocks of proteins. But there was more than a little incredulity in the audience. As they waited for Miller to appear, some scientists speculated among themselves about what Miller must have done wrong. Some guessed that he had been misled by contamination or had misinterpreted the results of his experiment. At the end of his presentation, Miller was bombarded with questions. Many of those were fielded by his doctoral adviser, the Nobel Prize–winning chemist Harold Urey, another key figure from the atomic bomb program. His presence lent the experiment the weight of his substantial scientific reputation.

The questions started to slow to a trickle as it dawned on the once-doubtful audience that they were indeed privy to an experiment that was sure to have historic impact. One of the last questions came from Fermi, a friend of Urey's since their days working on the Manhattan Project together. Fermi asked if this was indeed *the* way life had come about on the planet Earth or if it was simply *a* way that life *might* have appeared. "If God did not do it this way," answered Urey, "then he missed a good bet."

JUST TWO YEARS before the experiment that would make him one of the most famous scientists in America, Stanley Miller was still an undergraduate at the University of California at Berkeley. He might never have left, had it not been for lack of money. He had grown up just miles away in Oakland, and both his parents were UC alumni. His father, an assistant district attorney, had been a friend to Earl Warren, the future governor and Supreme Court justice, who lived nearby and whose children Miller often played with as a child. But his father passed away in 1946, and graduate school became out of the question unless Miller could secure a paid position as a teaching assistant. Only two schools made him an offer, and he ended up at the University of Chicago.

There, Miller found himself under the wing of the physicist Edward

Teller, part of the coterie of Manhattan Project scientists that found an academic home at the university. For the next year, under the guidance of Teller as his academic adviser, Miller turned to the question of the origin of the elements and how they might have initially formed within stars. In 1952, Teller left Chicago to head the secret American hydrogen bomb program. With Teller gone, Miller found himself without a mentor. His doctoral topic seemed no closer to completion than when he had begun the previous year, and he started to think about new subjects for his dissertation. He remembered a seminar he had attended about the early composition of the planets, in which Professor Urey had mused about the chemical composition of the Earth when life had first appeared.

Urey was an accomplished enough scientist that he stood out even in the university's star-studded chemistry department. His work on isotope separation and isolation of the heavy isotope of hydrogen, deuterium, had earned him a Nobel Prize in Chemistry in 1934. During the war, he had served as the director of the Special Alloyed Materials Laboratory at Columbia University, the branch of the Manhattan Project charged with enriching uranium for the first atomic bombs.

Urey was a chemist, but he never attached much importance to labels or specializations. His first degree was in zoology, and he initially hoped to study psychology. When he worked in the laboratory of the famous Danish chemist Niels Bohr, Bohr mistakenly assumed that Urey was a physicist. Ever since arriving in Chicago, Urey had found himself increasingly drawn to understanding the early development of the solar system and the origins of the planets.

At the lecture that inspired Miller, Urey had discussed the theories of Alexander Oparin, remarking offhandedly that it was almost amazing that nobody had seen fit to actually test what Oparin was proposing. The remark stuck with Miller, and he decided to approach Urey about taking up the challenge. This was a bit of a leap, since Miller had never been fond of experimental work. He had gravitated toward working with Teller because the professor shared his love of theory. Nonetheless, Miller would one day be remembered as having designed one of the most celebrated experiments of twentieth-century chemistry.

Urey agreed to take over as Miller's thesis adviser, but he balked at the young man's plan to hang progress toward his PhD on an experiment that Urey felt held perilously little prospect for success. Few had attempted to approach the problem of the origin of life experimentally since the days of the spontaneous-generation controversies of the late nineteenth century. Most capable scientists still saw the question of the origin of life as one not easily tackled by experiment *or* observation, making it something of an anomaly in the biological sciences. Chemistry was an *experimental* science, tending to focus on hands-on subjects: biological or geological phenomena well suited for careful observation. Pure theory was for physicists. The problem of the origin of life, on the other hand, was remarkably remote and obscure. Urey tried to steer Miller into a less ambitious course of study involving measuring the abundance of the element thallium in meteorites.

Yet Urey could also appreciate Miller's ambition. He often said that great scientists were great because they set themselves to tackling the big problems of science. A nuclear reaction, Urey liked to point out, was no more complex than any other kind of chemical reaction, only more important. He and Miller eventually reached a compromise. Miller would have a year to show results or he would switch his topic to the more achievable subject of thallium.

The two started to discuss how to approach the problem. They used Oparin's theory as a guide for how the first life might have appeared, adjusting it to Urey's theories of the early atmosphere's composition. Hydrogen, Urey figured, was the key. Hydrogen was by far the most common element in the solar system. At the time, one other scientist was pursuing a similar course of experiments into the origin of life, a Berkeley biochemist named Melvin Calvin. Calvin was one of the world's greatest authorities on the fabulously complex process of photosynthesis. He would be awarded a Nobel Prize in 1961 for explaining its underlying mechanism. Calvin assumed an early atmosphere consisting of carbon dioxide and water, energized by solar radiation. He simulated those conditions with an early particle accelerator, which had been invented at Berkeley by the nuclear scientist Ernest Lawrence. But Calvin's results were disappointing. His experiments produced trace amounts of the organic compounds formic acid and formaldehyde, but too little to be seen as significant and not really the types of organic material that could be easily understood as leading to the origin of life.

Urey, on the other hand, brought a unique perspective to the problem. His expertise was in the chemical composition of stars and planets, and his theories led him to hypothesize a somewhat different primitive atmospheric composition than the one Calvin tried to simulate. The biggest difference in their outlook was the relative abundance of oxygen. Urey thought it unlikely that the early atmosphere had contained any of it at all, aside from what was present in clouds. The composition of the Earth had been changed by unique circumstances, most notably the presence of living things, with their abilities to produce oxygen as a by-product of photosynthesis. Urey proposed that the early Earth instead had a reducing atmosphere, filled with hydrogen, methane, and ammonia.* The presence of methane was particularly important because it contained carbon, an element present in all living cells. Urey's hypothetical world was also turbulent, filled with frequent lightning storms. Since the eighteenth century, chemists had achieved remarkable results by subjecting gases to electric sparks. Electrical storms would have made the formation of organic materials easier.

Miller set about finding a way to construct a model of this early Earth. Working with the university glassblower, he began building a Pyrex version of Urey's primitive environment in a series of connected flasks and tubing. One flask was meant to represent the early ocean and contained water that could be heated to simulate evaporation. A second flask contained the gases methane, ammonia, and hydrogen, Urey's vision of the Earth's early atmosphere. An electric Tesla coil functioned as a lighting proxy. At the press of a button, Miller could launch a pale, blue-violet electric arc between two electrodes. The flasks were connected by a U-shaped tube with a condenser that would return any organic compounds made in the atmosphere flask to the experimental ocean, simulating rain in Miller's

* The presence of methane gas in Mars's atmosphere would be confirmed in 2009 by NASA scientist Michael Mumma. Like Urey, Mumma was raised in the United Brethren Church, which had tried to discourage his pursuit of science.

Stanley Miller and his "classical apparatus."

model Earth. Miller would go on to construct several models based on the design, one of which was capable of simulating volcanic heating. But his first simulation gained fame as one of the seminal experiments in the field of what would become known as "prebiotic chemistry." It was dubbed "the classical apparatus." Later, *Scientific American* published a do-it-yourself guide to re-creating Miller's experiment, and its replication became something of a right of passage among amateur chemists.*

* Author note from H. James Cleaves II: As a graduate student, I was lucky enough to have had Miller as my PhD adviser at UC San Diego. He was extraordinarily kind to those he worked with, but never shy in his criticisms of those he disagreed with. He was also fearless in the lab. I was asked to repeat Miller's experiment for an event celebrating its fiftieth anniversary. Miller had by that time suffered a stroke and was unable to explain the details of its execution. It took several weeks to figure out the details, and when the time came to flip the switch, Miller insisted on being present, along with some of his close friends. I was terribly worried some air might have entered the apparatus. My hope was to have everyone, including myself stand out in the hallway, connect a long extension cord to the Tesla coil, and flip the switch from a safe distance. Stanley would have none of it. I winced as I turned it on, fearing an explosion accompanied by shards of glass flying everywhere. Instead, I heard the faint

The experiment started showing promise almost immediately. After running it for one day, Miller noticed that the interior walls of his experimental Earth were coated with a yellow substance and the primitive "ocean" had turned a rich brownish red. After just two days, Miller was able to detect the presence of the organic compound glycine, an amino acid commonly found in proteins. When he finally shut the experiment down after six days, he conducted a thorough chemical analysis of what had taken place inside the confines of his miniature early Earth.

When word of Miller's experiment started to leak out into the small world of University of Chicago scientists, Urey was asked what he expected Miller to find. His answer was one word: "Beilstein." This was an allusion to *Beilstein's Handbook of Organic Chemistry*, ubiquitous in university libraries and containing detailed lists of all organic compounds known to date. In other words, Urey expected the apparatus to produce a little bit of everything. Instead, both men were astounded to see that a significant amount of the carbon present had been transformed into surprisingly few organic materials, including several types of amino acids. It was the best result they could have hoped for. Amino acids are the building blocks of the proteins that drive cellular metabolism. Most surprising, the amino acids that had formed in Miller's apparatus were the right *kinds* of amino acids—particularly glycine and alanine, commonly found in proteins. Alexander Oparin had hypothesized that proteins would have necessarily been the first cellular components to appear on the primitive Earth. Here was experimental evidence that seemed to confirm the first steps in the Oparin-Haldane hypothesis of how life originated.

The results seemed too perfect to be true. Urey insisted on a slow, meticulous process of confirmation and reconfirmation. In the 1950s, measuring amino acids was still a relatively primitive practice that involved blotting paper with the sample and applying various dyes to test for the presence

---

buzzing sound of the spark jumping between the electrodes. We then all got our faces up quite close to the flask and were mesmerized by little wisps of condensation swirling around the spark, looking something like fog tumbling down the hills on the San Francisco Peninsula when it rolls in late in the afternoon.

of various compounds. Eventually, Urey became convinced of the results, and it came time to publish.

By all rights, Urey was entitled to the lion's share of the credit. In most labs, the professor assumes most of the credit and graduate students can expect at best a second billing. But Urey was at the point in his career where he took more pleasure from watching his protégé's success than burnishing his own reputation. He understood that if they published the results under both of their names, Miller's contribution would be treated as an afterthought, so he insisted the results be published under Miller's name alone, a remarkably magnanimous gesture, considering the impact he knew the research was going to have.

Urey contacted the editors of the leading scientific journal in the United States, *Science*, and told them to expect the manuscript. But Urey's decision to leave off his own name—and the weight of his scientific authority—proved problematic. Such a monumental discovery, Urey supposed, would be harder to accept coming from a mere graduate student. After months passed without any word from the magazine, Urey complained to Howard Meyerhoff, chair of the editorial board, eventually asking that the manuscript be returned so that he could submit it to the less prestigious *Journal of the American Chemical Society*. Soon, Meyerhoff contacted Miller directly and assured him that the work would appear shortly.

The paper was published on May 15, 1953. Almost immediately, it made headlines throughout the world. An editorial in the *New York Times* described the Miller-Urey apparatus as "a laboratory earth. . . . It did not in the least resemble the pristine earth of two or three billion years ago; for it was made of glass." The *Times* went on to say that the experiment "made chemical history by taking the first step that may lead a century or so hence to the creation of something chemically like beefsteak or the white of an egg." *Time* magazine reported that Miller and Urey had proved "that complex organic compounds found in living matter can be formed. . . . If their apparatus had been as big as the ocean, and if it had worked for a million years instead of one week, it might have created something like the first living molecule." Alexander Oparin, reading the news in Moscow, doubted that the experiment could be true. Almost overnight, Miller became one of the most famous scientists in the world, and certainly the most famous American ever to have taken on the question of the origin of life.

IN OCTOBER OF 1957, a pair of Central Intelligence Agency officials stopped by Miller's office. They were interested in a letter that Miller had recently received from Alexander Oparin. It was an invitation to a symposium that the Russian was hosting in Moscow on the origin of life, which Miller had accepted. In the four years since Miller's experiment, the subject had been reinvigorated in the scientific community. Scores of young physicists and chemists were drawn to the questions Miller and Urey had raised. And although he was just twenty-seven years old, Miller had already laid claim to the title of "godfather of prebiotic chemistry."

For months, Oparin's invitation had posed a dilemma for Miller. It had come during the height of the Cold War, just a year before the Berlin crisis. Joseph McCarthy had died earlier that year, falling victim to years of chronic alcoholism, but the climate of fear that McCarthyism had engendered lingered. This was particularly true in academia. The persecution of Robert Oppenheimer, the former head of the Manhattan Project, stood as a warning of what could happen if certain lines were crossed. The testimony against Oppenheimer given by Miller's former professor, Edward Teller, had led to Teller's ostracism in American academia.

A visit to Moscow was not a step Miller could take lightly. He had written to Harold Urey asking for advice. Urey was one of the many nuclear scientists to face scrutiny during the McCarthy hearings. He had argued on behalf of Julius and Ethel Rosenberg before their 1953 execution for stealing nuclear secrets on behalf of the Soviets. Later, Urey himself had been called before the House Un-American Activities Committee. Urey told Miller he had to make up his own mind but cautioned, "One never knows what a McCarthy will do in the future. It is a very sad situation."

Miller decided to accept Oparin's invitation. The visit from the CIA officers came shortly thereafter; they were visiting all the scientists headed to Moscow for the conference. Some in the upper levels of the US govern-

ment feared that the Soviets might lay claim to creating life in a laboratory. Such a discovery would have represented a significant propaganda
victory for the Soviet regime, and the Americans could ill afford any more
of those. Earlier that month, the USSR had launched the first artificial satellite, Sputnik-1. Across the United States, amateur radio enthusiasts tuned
in on their ham sets to hear the eerie cricket-like chirping of Sputnik's
radio transmitter as it hurtled, untouchable, over American airspace. The
sound was quickly taped and rebroadcast on commercial airwaves to a
shocked public. In December, the first American satellite was launched in
response, only to explode ignominiously in front of a live national television audience. Newspaper headlines the next day bemoaned the failure of
America's "Stayputnik." There was a growing perception that the USSR
was gaining the upper hand in the scientific front of the Cold War.

Editorial cartoon appearing in the *Washington Post*, December 31, 1956.

The CIA agents wanted Miller to report back to them on the origin-of-life work being done in Moscow. Miller agreed but was flabbergasted by the agents' inability to understand even the most basic facts about the subject they were investigating. When he returned, they spent most of their time debriefing him on the lack of air-conditioning in Moscow and the personalities of the scientists involved.

The origin of life was still a politically sensitive subject in the United States. It had been only three decades since the Scopes Monkey Trial in Tennessee, where three-time Democratic presidential candidate William Jennings Bryan had railed against the inability of science to explain the origin of life. After the results of the Miller-Urey experiment made news headlines, a Gallup poll asked whether it was possible "to create life in a test tube." Only 9 percent of respondents thought it was. But Cold War competition between the superpowers trumped any domestic political sensitivities. The United States was about to embark on the most ambitious and far-reaching government-funded science program since the Manhattan Project, and the scientists studying the origin of life were going to play a major role.

LATER IN THE WINTER of 1957, a young Fulbright scholar from the University of Wisconsin named Joshua Lederberg stopped in for a visit with Haldane at the elder scientist's home in Kolkata. Lederberg was just thirty-two years old, but he had already built an impressive reputation in the field of genetics for his discovery that bacteria could mate and exchange genetic information. Previously, bacteria were thought to simply pass genetic information to their offspring completely unchanged, creating exact copies of themselves. Lederberg had shown that their reproduction was much more complex. In 1958, Lederberg's work on bacterial genetics would earn him the Nobel Prize in Physiology or Medicine.

Lederberg was one of those to receive an invitation to Oparin's Moscow symposium. Ultimately, he had decided not to attend, opting to continue his work at the virologist MacFarlane Burnet's laboratory in Melbourne, Australia. On his way home to Wisconsin, Lederberg decided to make a stop in India to pay his respects to Haldane, who had only just begun his self-imposed exile from Great Britain. There was a lunar eclipse that night, an event of great significance in Hinduism, and Lederberg had to make his way through streets crowded with religious revelers to get to Haldane's home. Naturally, the two men's conversation veered to the stars, and to the topic that was on almost everyone's minds: the USSR's launch of Sputnik. It was the fortieth anniversary of the October Revolution, and Haldane worried that the Soviets would try something daring. He wondered if they might try to make a show of their newfound military capabilities by exploding a nuclear weapon on the moon—a demonstration that would be visible on Earth. The men spent the evening lamenting the thought that the promise of space exploration was being lost to a game of geopolitical one-upmanship between the superpowers.

When he arrived back in the United States, Lederberg immediately set about trying to establish a toehold in the nascent American space program. Within a month, he had circulated two memos throughout the National Academy of Sciences discussing the potential for a new "cosmic microbiology" and "lunar biology." Lederberg was advocating that the scientific search for the origin of life be extended to the space program as a search for life on other planets. Finding life in space would have important implications for the scientists trying to discover the origin of life on Earth. As a bacteriologist, Lederberg posed his suggestions as a matter of national security. He worried that the first life human beings would encounter in space would be bacterial and potentially very dangerous, capable of wreaking devastation much the way the introduction of bacteria from Eurasia had once devastated native populations in the wake of Columbus's voyage to America, an idea that Lederberg popularized in an article entitled "Moondust" that he wrote for *Science*.

Lederberg's suggestions grabbed the attention of Hugh Dryden, who was emerging as one of the most important figures in America's rapidly expanding space program. In July of 1958, President Eisenhower signed the National Aeronautics and Space Act, creating NASA. Dryden was named the deputy administrator, and one of his first acts was to set up a Space Sciences Board to advise the new agency. Lederberg was named to head a panel on extraterrestrial life. From his new position, and with the enormous funding of the space program to draw upon, Lederberg began to attract some of the leading names in origin-of-life research, such as Harold Urey, Berkeley's Melvin Calvin, and Stanley Miller, who was already speculating about the possible presence of life on other planets. Lederberg also began snatching up some of the most promising young minds in the field.

One of the men Lederberg recruited was a young astrophysicist named Carl Sagan, who had been a student of Urey's at the University of Chicago when Miller conducted his electrical discharge experiment. Sagan quickly became one of the most enthusiastic embracers of the space program, and he carried his enthusiasm for understanding the origin of life with him when he left Chicago. His unique knack for popularizing science was apparent almost from the beginning—something that Lederberg and his coterie of scientists would benefit from in the years to come.

By 1959, the term "exobiology" had begun to appear in Lederberg's private letters to describe the way in which the search for the origin of life on Earth could be used to hunt for life in outer space. The term caught on quickly. But at its root, the work always remained focused, ultimately, on the search for the origin of life on Earth. As Carl Sagan later wrote, exobiology was nothing more than "extending [Stanley] Miller's results to astronomy."

At the start of the 1950s, research into the origin of life was pitifully underfunded and neglected at universities. The Miller-Urey experiment had come about only by what Miller later called "bootlegging" funds marked for other research. They staged the whole experiment for less than a thousand dollars. But by the early 1960s, research into the origin of life had begun to draw on the seemingly bottomless pockets of the American space program. As early as 1959, money began to pour in for work on instruments to detect life on other worlds. Twenty years after it was created, NASA was easily the world's largest funder of origin-of-life research in the world.

One of the first to receive a grant was Wolf Vishniac, a microbiologist and exobiologist at Yale Medical School, who received funding for a device to detect microorganisms present in the soil of other planets. He named his device the "Wolf Trap." In the coming decades, scientists from the exobiology program would be instrumental in the *Apollo* missions to the moon, and central to the *Viking* missions to Mars. They went on to produce important new theories, such as the Gaia hypothesis and the grim potential climatic effects of a "nuclear winter" resulting from atomic warfare. And as they searched for signs of life on other planets, they continued to take important steps toward understanding just how life arose on Earth.

That understanding was about to change in some very fundamental ways. Just a few weeks after the results of the Miller-Urey experiment were published, a team of scientists in Great Britain would tease apart the molecular structure of DNA, a discovery that, in the years to come, was going to upend everything that scientists believed about the mechanisms of biological inheritance. The search for the origin of life—and even the most basic understanding of how life was constructed—was about to undergo a major revolution. Much of it would play out among the scientists affiliated with the NASA exobiology program.

THE NUCLEIC ACID MONOPOLY

All of today's DNA, strung through all the cells of the earth, is simply an extension and elaboration of [the] first molecule.

—LEWIS THOMAS, *The Medusa and the Snail*, 1969

N EIL ARMSTRONG SAT in the command module of the *Apollo 11* spacecraft and stared out the window at the surface of the moon. It had been three days since the mission launched from Earth, and the ship was now settled in orbit 60 miles above the moon, awaiting the moment when Armstrong and Buzz Aldrin would enter the landing module *Eagle* and begin their descent to the first extraterrestrial surface upon which a human being would ever stand.

Below them lay a vast, bluish-tinted basin filled with hardened lava formed by ancient volcanic eruptions. It was the *Mare Tranquillitatis*, so named by the seventeenth-century Italian Jesuits Francesco Grimaldi and Giovanni Riccioli. The first map of it had appeared in Riccioli's great almanac of astronomy, the *Almagestum novum*, in 1661. Misled by its color, the Italians had mistaken the basin for a sea.

The *Apollo 11* astronauts simply called the *Mare Tranquillitatis* by its English translation, the "Sea of Tranquility." When they returned from the moon, they planned to bring a sample of it back with them to be studied by geologists and life scientists from NASA's exobiology program. For the American public, simply landing on the moon would be enough. But for NASA scientists, especially the origin-of-life scientists in the exobiology program, the prospect of being able to study pieces of the moon was immeasurably enticing.

Apollo 11 was the fifth manned space flight undertaken by the US space program. The previous mission, *Apollo 10*, had been a dress rehearsal for the lunar landing that Armstrong and his fellow astronauts were about to attempt. Lunar probes launched from *Apollo 10* had taken detailed photographs of potential landing sites. NASA scientists had combed over the pictures, searching for the ideal landing spot for the next mission. *Apollo 11* would be following an orbit that was roughly in line with the moon's equator, so the landing spot had to be near enough to its trajectory that the landing craft would have sufficient fuel to make its descent. But they also wanted to choose a site with a wealth of geological features. Armstrong and Aldrin would have precious little time on the surface, and they were about to undertake what would amount to the most important geological survey in history. Even if everything went like clockwork, they would have just a little over two hours to gather as much of the moon's geological diversity as they could.

Evidence of past volcanic activity was enticing to the mission planners for the same reasons it had been enticing to men like Charles Darwin. The lava could preserve things that wouldn't otherwise be found in normal rock formations. But the same volcanic formations that were so attractive to scientists back on Earth presented a challenge to the astronauts. One of the biggest dangers posed by the mission would be planting a fragile landing craft safely on the surface. The Sea of Tranquility represented a compromise. Despite its geological promise, the region wasn't overwhelmingly mountainous, making it a relatively attractive site for the astronauts to land.

Armstrong spent most of the trip to the moon poring over maps of the region. Now that they were comfortably in orbit, he could make out its features with his own eyes. With the moon now standing directly between *Columbia* and the sun, the landscape was bathed in a blue glow of light reflected off the Earth. Craters were clearly defined, almost three-dimensional in the earthshine. Aldrin was the first to make out the 3-mile-wide crater that marked their landing site. It looked rugged and ill suited for landing, as if the planners back at mission control, in their desire to acquire a good geological haul, had been too daring. But as the area gradually came under the direct light of the sun, it began to look less foreboding.

The next day, Armstrong and Aldrin climbed aboard the *Eagle*, leaving the third member of the crew, Michael Collins, behind in command of *Columbia*. Their descent from orbit was problematic. The computer's alarms sounded twice—the result of a hardware malfunction—and Armstrong had to take early manual control of the landing. As they flew closer to the site, they were dismayed to see that the area was strewn with boulders. That would no doubt please the geologists on Earth, but Armstrong knew it would make for a tricky landing. Nonetheless, he managed to skillfully guided the *Eagle* just 350 feet above a large cluster of rocks, touching down near a crater the size of a football field, big enough to pose an obstacle but too small to have been spotted on the maps made by *Apollo 10*. NASA personnel back at mission control in Houston waited in silence, knowing that the astronauts were in the middle of the most dangerous phase of the mission. Then they heard Neil Armstrong speak the first words ever spoken on the moon: "Houston, Tranquility Base here. The *Eagle* has landed."

After a rest of about 2 hours, Armstrong and Aldrin began suiting up in the bulky space suits that had been designed for their moonwalk. Armstrong then descended down a ladder, flipping on a camera that had been mounted on the side of the ship. As he reached the bottom rung, he began to examine the surface. In training, he had been grilled to describe everything for the benefit of the scientists back at NASA. "The surface appears to be very, very fine grained as you get close to it," he said. "It's almost like a powder."

He hopped onto the surface and took a few steps. Some 240,000 miles away, most of America and much of the world sat breathlessly watching on television or listening on radios.* While sitting in the *Eagle* shortly after its touchdown, Armstrong had thought of what he would say when the historic moment arrived: "That's one small step for man, one giant leap for mankind."

Without his suit, Armstrong would have died in seconds. Yet he was struck by the serenity of the scene that confronted him. The rising sun bathed the moonscape in bright light. It looked different from the pictures he had seen from the probes, and the lack of an atmosphere gave every-thing a clarity he had never experienced on the Earth. It was beautiful, if stark. Aldrin called it "magnificent desolation."

Armstrong began snapping pictures from a camera built into his space suit, but stopped when he was interrupted by a voice from mission control urging him to get on with his real work. They were in the middle of the most important geological survey in history, and it would have to be com-pleted in the roughly two and a half hours the astronauts had before their supply of oxygen began to run out.

From a pocket in his space suit, Armstrong took out a collapsible rod with a bag at one end. He began scooping dust and a couple of small rocks from the ground below him, until the sample bag was packed full of gray-black powder. Even if something went awry and the moonwalk had to be aborted, the astronauts would at least return with this bag of what the scientists back at mission control called the "contingency sample." After a series of other experiments, the astro-nauts went about a more discerning process of collecting what would be called the "documented sample." A phone call from President Nixon had put them behind schedule, and they had to rush through what was supposed to be one of the most careful and deliberate phases of their excavation. Armstrong set about filling two more aluminum containers that the scientists had dubbed "rock boxes." While Aldrin began hammering away at the surface with a core tube searching for a specimen that would provide scientists a picture of what lay just under

---

* An estimated one out of every five people on Earth followed the broadcast.

Buzz Aldrin drives a core tube sampler into the lunar soil. Photographed by Neil Armstrong.

the moon dust that represented a kind of topsoil, Armstrong hurried about with a long pair of tongs, grabbing the rocks that looked most interesting.

THE ASTRONAUTS RETURNED TO Earth with 45.5 pounds of the lunar surface for the scientists at home, but there would be a delay before the business of serious study could begin. From the outset of the *Apollo* program, NASA officials had taken seriously Joshua Lederberg's warnings of contamination from lunar bacteria. An elaborate quarantine center had been built at the Johnson Space Center in Houston, with a containment room for the lunar samples just down the way from the one that had been built for the astronauts themselves. In the case of the moon rocks, scientists had more to consider than the dangers of a lunar organism: contamination of the moon rocks themselves from microorganisms on Earth was a major concern, as it would taint the precious samples and could lead to misleading results.

Three weeks after the *Eagle* lifted off from the moon, the samples were ready to be parceled out for study by four research groups in NASA's exobiology program, each led by scientists who either had been or would be key figures in the search for the origin of life. Two of the teams were headed by scientists directly employed by NASA—one headed by geologist David McKay, who had personally trained Armstrong and Aldrin for their scientific work during their moonwalk; the other, by Sri Lankan chemist Cyril Ponnamperuma, who had become an important authority on the origin of life in NASA exobiology circles. A third box of samples was shipped off to UC San Diego, where they would be studied by a group led by Harold Urey, whose theories on the composition of the lunar environment had by then earned him the title "father of lunar science." The fourth box of samples was delivered to the University of Miami, where it would be examined by a team headed by a colorful iconoclast named Sidney Fox, a six-foot-four chemist with a reputation as an absentminded professor, capable of spending hours in parking lots searching for his car and falling asleep in midsentence while standing up and delivering a lecture.

Fox's life story was almost as colorful as his personality. His father was a wig maker; his mother, a Ukrainian Jew who had fled tsarist Russia when she was just eleven years old, stowed away in a crate on board a steamship. Fox grew up in Los Angeles with a passion for music, particularly Benny Goodman–style big-band jazz and Broadway musicals. In his twenties, he dabbled in composing. While studying chemistry at UCLA, he even wrote the scores for several of the university's well-received annual musical revues. In 1935, he received a phone call from Walt Disney Studios asking whether he would be interested in composing the score for a movie based on an old Brothers Grimm fairy tale. It was going to be called *Snow White and the Seven Dwarfs.*

Fox was enthralled, but first he sought out the advice of his mentor at UCLA, a professor named Max Dunn. "You are going to make a choice between music and chemistry," said Dunn. "And it is going to be chemistry."

After UCLA, Fox moved to the California Institute of Technology. Founded in 1891 as a vocational school, tiny Caltech had, in a few decades, transformed itself into a world-class scientific research institution. Despite its relatively small size, as of 2015, Caltech scientists had won thirty-four Nobel Prizes, and the university had the distinction of having the highest faculty citation rate in the world. Even by the time Fox arrived in the 1930s, it had attracted some of the most important scientific minds in the United States to its faculty, including two who were making remarkable strides toward understanding the way living things work on a subcellular level: the chemist Linus Pauling and the biologist Thomas Hunt Morgan. Pauling was a pioneer in quantum chemistry and, in later years, would become one of the key elaborators of the molecular structures that make up living cells. Morgan had won fame as an evolutionary biologist who had just won a Nobel Prize for discovering, in fruit flies, the precise role that chromosomes play in genetic inheritance. Both men exerted a strong influence on Fox and the shape of his future scientific work.

At Caltech, Fox began to take a keen interest in evolution, particularly the prebiological history that resulted in what he often called the spontaneous appearance of the first life-forms. Fox was ambitious. He wanted to work in a field that would allow him to make a real impact. To Fox, the question of the origin of life was the central biological problem, precisely the sort of question that so many other scientists avoided and nobody seemed close to being able to answer. It was also a question that enabled him to forge new ground and challenge long-held assumptions of the scientists who came before him. Ironically, toward the end of his career, many scientists would end up accusing Fox of being stuck in scientific assumptions that were losing relevance to the understanding of the earliest life-forms.

AS A PROFESSOR AT Florida State University, Sidney Fox had become one of the more prominent origin-of-life scientists in the United States. He was also one of the first to be drawn into the web of NASA's exobiology program, which was rapidly expanding along with the rest of the agency. He had been asked by NASA to organize the first American scientific conference on the origin of life, the Wakulla Springs conference, where he had brought Haldane and Oparin together. Ever since, he had aggressively tapped into NASA funding, using it to establish the first exobiology lab in the country, the Institute for Space Biosciences at Florida State. When Fox moved on to the University of Miami in 1964, NASA funding helped him establish a freestanding research facility, the Institute of Molecular Evolution, where several luminaries in origin-of-life research would be trained over the next two decades.

Prior to the *Apollo 11* landing, most NASA scientists predicted they would find a bounty of organic compounds on the moon. Cyril Ponnamperuma had been particularly confident. As it turned out, the *Apollo* astronauts found a desolate landscape without a protective atmosphere provided by a planet like the Earth and shaped by billions of years of exposure to the sun. Though the lunar samples from *Apollo 11* were eventually found to contain small traces of amino acids, these were so scarce that it was hard to see them as significant. As Fox would later write, it was as if the surface of the moon had been "baked to a cinder."

Fox wasn't terribly disappointed. Officially, his role in the space program was to look for traces of organic compounds, evidence that the precursors of life were strewn throughout the solar system. But his real interest in exobiology wasn't so much the possibility of life in space as it was what space could tell us about life on Earth. He was looking for clues to how life might have begun on Earth, as he put it, "to test proposed concepts of steps leading to the emergence of life." By the time of the *Apollo 11* mission, Fox had already come to believe that his laboratory experiments had shown the next crucial step in the evolution from amino acids to full-fledged proteins, and that he had solved the riddle of what Haldane would have called the "half-living" stage in the evolution of a living cell might have been like.

From the beginning of his experimental work on proteins, Fox had had his share of detractors. By the time of *Apollo 11*, those critics had grown numerous. Scientific understanding of the molecular composition of living cells had grown exponentially in the decade and a half since the famous Miller-Urey experiment, and those advances were undermining Fox's conception of the earliest life-forms.

Like almost everyone in the field by then, Fox believed that a full-fledged living cell had not simply appeared fully formed on the primitive Earth. It had become apparent that simple *components* of living cells must have arisen first, which could then begin the long evolutionary process that would produce cells as we know them today. But which crucial component had it been? In many ways, this was the same problem faced by Fox and other scientists in the NASA exobiology program as they looked for microscopic life on other worlds: what was it about living cells that actually made them *alive*?

IN 1944, the Austrian theoretical physicist Erwin Schrödinger wrote a book called *What is Life?* The question was an old one. Even the most primitive ancient peoples noticed the differences between plants and animals and the inanimate world about them. The vitalists of earlier centuries had been captivated by the question of what exactly that difference was, or at least what caused it. But few had ever sought to really quantify the problem as Schrödinger had, and his book generated a great deal of excitement in the scientific world.

Schrödinger was a physicist, and a very successful one. His elaborations of quantum mechanics would eventually win him a Nobel Prize, and he approached the phenomenon of life as one might expect a physicist to. For him, the basic element inherent to all life was its ability to avoid the inevitable fate of all matter in the physical world: the decay into entropic chaos. A living thing does this by what Schrödinger described as "drinking order" from its environment: drawing in chemical elements and energy from the environment, and then transforming and rearranging them, via a functioning metabolism. But Schrödinger also singled out another factor in what made living things living: mutation, the replication with change that lies at the heart of the modern concept of evolution.

Schrödinger deduced that the basis of genetic inheritance must lie in a molecule he called an aperiodic crystal. He chose a crystal because he believed that this molecule had to be ordered and stable, and thus able to persist over generations, which would not be true for a coacervate-like suspension. The crystal had to be aperiodic because he believed it would have to be able to store limitlessly variable information to allow for mutation and evolution. In other words, it had to be a single molecule structured in a way that atoms could store information.

Since the complexity of the first living thing must have stopped well short of a full-fledged cell, it stood to reason that some *part* of a cell had come first. The first half-living thing—what would come to be called a protocell—would have to have been able to do both of two things: metabolize using components of its environment and replicate with modification. Metabolism and replication were the same factors that Oparin and Haldane had singled out in their own hypotheses for the origin of life. The problem that would eventually arise was that these two functions are handled by different but mutually interdependent subsystems within the cell itself.

In later years, it would usually be described as the chicken-or-the-egg paradox of the origin of life. But in the mid-twentieth century, when Miller and Urey had reinvigorated the field with their experiment at the University of Chicago, it was a problem that hadn't yet reared its head. Though the workings of enzymes that govern metabolism were becoming well understood, little was concretely known about chromosomes. Since Thomas Hunt Morgan's elucidation of the chromosome's role in genetic inheritance, it had been clear that chromosomes play a central role in the functioning of genetics. But no one yet understood what they were actually made of. It was reasonable to assume that replication was handled by the same part of the cell that governed metabolism, that the chicken and the egg were the same thing. The reality was that scientists didn't yet really know all that much about cells.

THE CELL PROVIDES one of the strongest pieces of evidence of the deep evolutionary connection between all life on Earth. Just as Geoffroy Saint-Hilaire had seen the similarities in such seemingly different appendages as a bird wing and a human hand, microbiology, as it became more sophisticated, revealed equally compelling similarities in the structures of living cells of diverse organisms. The structures, the functions, even the language of genes, are too similar for cells not to have originated from the same cellular ancestor.

There are only two basic types of living cells: prokaryotes (from the Greek for "prenucleus") and eukaryotes (those containing a "true nucleus"). Single celled and lacking a nucleus, the prokaryotes are the simplest organisms. All multicellular organisms—plants, animals, fungi—are eukaryotic. Every multicellular species is like a colony of cells, each of which is programmed to perform specialized tasks and depends on the functions of other cells for its survival. The human body has so many cells that it is difficult to say how many there are. Some estimates have put the number as high as 100 trillion, but most guesses are about one-third that number.

The first person to actually see a living cell was Antonie van Leeuwenhoek, although it is his seventeenth-century contemporary Robert Hooke, who usually receives credit. In his book *Micrographia*, Hooke described microscopic structures that made up a piece of cork. These were not actually cells, but the remains of cell walls, made up of leftover cellulose and lignin. The name he used to describe them—"cell," a derivation of the Latin word *cella*, for "small room"—stuck. Thus, cells were named after the part that was easiest for the first microscopists to observe, the protective lipid layer known as the cell membrane.

At its most fundamental level, a cell is nothing more than a few basic elements. But the way these elements are organized is enormously complex, coming together to form a dynamic factory with complex machinery, most of which is made up of proteins. There may be tens of thousands of types of proteins, each carrying out a specific task in the cellular assembly line. Pretty much everything living organisms do—respiring, eating, growing, reproducing—is accomplished by or with their help.

In some ways, the cytoplasm, the viscous protein and nucleic acid solution housed within the confines of the cell membrane, came to be seen by scientists as a modern version of the protoplasm, the enigmatic essence of living things that gave them all their lifelike properties. Proteins were seen as particularly important. In the first half of the twentieth century, it was easy to assume that proteins carried genetic information as well. The importance of proteins in cellular metabolism was well established, although some scientists were beginning to have doubts about their role in heredity.

LIKE MOST SCIENTISTS who were focused on the origin-of-life question at the time, Sidney Fox was firmly in the camp of those for whom the protein reigned supreme. At Caltech, Morgan, one of the fathers of modern genetics, frequently told him, "Fox, all the problems of life are problems of proteins." That message was reinforced after Fox left Caltech for a position in the lab of one of the world's greatest authorities on proteins, the chemist Max Bergmann, a German Jew who had fled Nazi Germany and established a laboratory at Rockefeller University in New York.

The start of World War II sidetracked Fox for a while. He returned to California, joining the war effort at a private lab where he helped invent a process to isolate from shark livers vitamin A that could be used by Allied pilots to enhance their night vision. After the 1953 Miller-Urey experiment had generated so much excitement in the scientific world, Fox returned to the "big question" that had intrigued him since his days at Caltech, the question of the origin of life.

The Miller-Urey experiment triggered a rash of similar experiments seeking to duplicate the formation of amino acids. Many involved modifications of Urey's vision of the early-Earth environment. Some used different gas combinations to simulate the early atmosphere. Various energy sources were substituted for the lighting Miller had used. But Fox wasn't interested in retreading the same ground Urey and Miller had walked. The question of how amino acids had appeared on Earth may not have been firmly settled, but it appeared to be nearly so. The origin of organic molecules no longer felt like an insurmountable mystery, and certainly not the obstacle it might once have seemed. Instead, Fox turned to the next step in the development of living organisms. He wanted to understand how amino acids could form some type of early prototype of the cell that would have represented the "half-living" stage Haldane had once referred to. For Fox, that next step had to have been the formation of a protein, or something close to it, from amino acids.

This was a more difficult problem than Miller and Urey had confronted. Even the smallest of proteins contain a long string, or polymer, of amino acids that have to be arranged in a precise sequence. Thus, the term "sequencing" in proteins refers to establishing the exact order in which the amino acids are arranged. Later, the term "sequencing" became more commonly associated with the search for the order and placement of genes in chromosomes, and for the order of individual nucleotides within genes.

To get from the simple amino acids Miller and Urey had found in their experiment to full-fledged proteins was an exceedingly complicated proposition. Yet Fox seemed to stumble upon what he saw as the answer rather quickly. After the Miller-Urey experiment had invigorated the study of the origin of life in academic circles, Fox turned in earnest to the problem of how simple amino acids could be built into more complex structures. While away from the lab giving a series of lectures, he had an epiphany: what would happen, he wondered, if amino acids were evaporated in an environment resembling Darwin's warm little pond?

Fox and his research team found that purified amino acids heated to 175°C organized themselves in a seemingly nonrandom way into chains of amino acids similar to protein polymers. These strings of amino acids were short compared to real proteins, but they did exhibit signs of catalytic activity similar to that of proteins. Convinced he had discovered the next step on the road to life from nonlife, Fox named these short amino acid strings *proteinoids*. By 1959, he had found that when his dry proteinoids were reintroduced to hot water, they spontaneously formed tiny microspheres that he increasingly described as "lifelike." In May of 1959, Fox announced his discovery in a paper published in *Science*. His experiment, he claimed, was the basis of what he called "a comprehensive theory of the spontaneous origin of life at moderately elevated temperatures."

As his study of proteinoid microspheres continued, Fox became more and more convinced that he had discovered the precursors of modern cells. They had an outer shell that in some ways mimicked a cell membrane, such as by being selectively permeable to certain biological substances. They acted as catalysts, speeding up chemical reactions much the way proteins do. The microspheres were even capable of absorbing other microspheres, allowing them to grow, bud, and divide into new microspheres. Later, he would discuss his microspheres as if they were alive in some primitive way.

Many scientists were skeptical from the outset. Miller was one of the most prominent of these critics. In a letter published in *Science* following Fox's first claims, Miller and Urey questioned the plausibility of the geological conditions that would have been necessary for Fox's proposed scenario. Even in a laboratory setting, proteinoids required a precise series of very specific steps, heating and cooling coupled with drying and rehydration. Fox hypothesized a tidal basin next to a volcano as the setting where the geochemical creation of proteinoids might have taken place. Miller and Urey doubted that such steps could have occurred on the early Earth and downplayed the idea that a volcano would have been a conducive place for life to arise. Fox, they argued, had discovered a fascinating chemical phenomenon, but not a phenomenon relevant to the origin of life. Fox's proteinoids were certainly not alive, nor were they capable of evolving into something that was alive.

It was this last criticism, coupled with the burgeoning science of molecular biology, that eroded the significance of Fox's work in the eyes of most scientists. In the years following the Miller-Urey experiment, scientists had acquired a greater understanding of the roles of the complex molecules found in biological cells—an understanding that was steadily undermining the idea of the protein being key to every life process. Proteins were gradually losing their preeminence among those trying to understand the origin of life, and the search for the ancestor of the modern cell was slowly but surely drifting toward other cellular components. The focus was steadily turning to nucleic acids, long misunderstood and underappreciated.

On a spring day in 1953, a tall, sandy-haired young man with a bit of a stoop walked into the Eagle Pub in Cambridge and boasted loudly that he and his companion had just discovered the secret of life. It was a bold claim, considering it was being made by a scientist who had not yet completed his PhD. But then, Francis Crick was never shy about sharing his high opinion of his own abilities. James Watson, the man who shared pints with Crick at the Eagle Pub and later a Nobel Prize in Stockholm, would eventually write a book about the discovery of the structure of DNA titled *The Double Helix*. The book's first line was, "I have never seen Francis Crick in a modest mood."

It would take the rest of the world another decade to appreciate just how important Crick and Watson's discovery truly was. Eventually, though, their discovery of the structure of DNA would indeed come to be seen as one of the most important scientific advances of the century, and Crick's would become one of the most famous names in science.

A SHOEMAKER'S SON from Northampton, then the shoemaking capital of Great Britain, Crick was old for a doctoral student. And although his grandfather, Walter Drawbridge Crick, was, in his spare time, a naturalist and an acquaintance of Charles Darwin, and even coauthored Darwin's last submission to *Nature*, Crick had scant experience in biology. He spent his undergraduate years at University College London, where Robert Grant had once instructed Henry Charlton Bastian. Like Bastian, Crick was most intrigued by two important but little-understood biological problems—the phenomenon of consciousness and the origin of life—yet he had settled on what to him was the largely unsatisfying pursuit of a degree in physics. When World War II broke out, Crick had been working on what he later described as the "dullest problem imaginable," the measurement of the viscosity of water at temperatures between 100°C and 150°C. He was rescued from this fate when a German airplane dropped a sea mine on the laboratory where he worked and destroyed his experimental apparatus.

Crick then found himself designing sea mines of his own for the British Navy's Admiralty Research Laboratory. He came up with a rather ingenious design for a mine that could be triggered only by extreme magnetic fields of the sort used by German minesweepers.

At the end of the war, Crick was still a student in search of a PhD.

Like a generation of young physicists inspired by Schrödinger's book *What is Life?*, Crick was determined to move into biology. In 1949, he found a research position at Cambridge's Cavendish Laboratory. "The Cavendish," as it was known, was the most prestigious physics laboratory in Great Britain, as well as the site of some of the most revolutionary discoveries that were unlocking the secrets of the inner workings of cells.

In 1912, not long after the discovery that X-rays were, in fact, waves, a twenty-five-year-old Cambridge student named William Lawrence Bragg had hit upon the idea of using the diffraction patterns of X-rays to obtain a picture of the arrangement of atoms of crystals. Soon, scientists had found ways to crystallize isolated samples of the components of living cells, enabling them to get a glimpse at their atomic-level structure and learn the intricate details of how they actually worked. The discovery would lead to a revolution in the understanding of biochemistry and turn Bragg into history's youngest Nobel laureate.*

By the time Crick arrived, Bragg was running the Cavendish. Under his leadership, the Cavendish had become the center of some of the most advanced crystallography research being conducted in the world. Most of the work centered on proteins, and Crick was initially assigned the task of critiquing the research of one of Cavendish's brightest stars, the Austrian microbiologist Max Perutz, who was diligently trying to establish the molecular structure of hemoglobin. Perutz hoped that the structure of that protein would unlock the secret of how genes are transmitted, but Crick found himself unconvinced by Perutz's hypothesis. Crick had begun to lean toward the view that the secret to genetic inheritance lay elsewhere in the cell, in the oft-overlooked nucleic acid called DNA.

Back in 1871, a Swiss-German chemist named Friedrich Miescher had isolated a new biochemical substance from cells that he had extracted from the pus-soaked bandages obtained from a nearby hospital. Miescher was puzzled that the substance contained nitrogen and phosphorus but not the sulfur that would have established it as a protein. Because the substance had been drawn from the nucleus of cells,

* Bragg's record as youngest Nobel laureate was finally broken in 2014 by Malala Yousafzai, who won the Nobel Peace Prize at the age of seventeen.

he named it "nuclein," which was eventually changed to "deoxyribonu-cleic acid," or simply DNA.* But nobody really knew exactly what DNA did or that it had any relationship at all to the purely theoretical things known as genes.

More than half a century after Miescher's discovery, the world was treated to a vital clue that DNA played a more important role in genetic inheritance than anyone had previously thought. In 1943, a Canadian-born physician named Oswald Avery began a series of experiments on bacteria at the Rockefeller Institute in New York. Chemical analysis had by then become sophisticated enough to show that bacterial cells were composed of both proteins and nucleic acids, and that the two components could be separated. Working with a bacterium that had been shown to cause pneumonia, Avery found he could change harmless bacterial strains into virulent ones simply by exposing them to the DNA of pathogenic strains. The lesson was that DNA by itself—completely free of the presence of pro-teins—had the power to pass on genetic traits.

Joshua Lederberg would one day call the experiment "the histori-cal platform of modern DNA research," but it took a great deal of time for most scientists to accept its full significance. Everybody knew about Avery's findings, but too many great minds had invested themselves in the centrality of protein. The idea of DNA as the carrier of genetic infor-mation represented a fundamental paradigm shift in the understanding of biochemistry and the working of cells. There was a great deal of resis-tance in the scientific community, even within the Rockefeller Institute. As late as 1951, in an essay marking the half-century anniversary of the rediscovery of Mendel, the great geneticist Hermann Muller—the first to realize that genes were subject to mutations—would write, "We have as yet no actual knowledge of the mechanism underlying that unique property which makes a gene a gene—its ability to cause the synthesis of another structure like itself."

For a growing number of scientists, however, the implication of Avery's experiment was clear: DNA was indeed the central agent of genetics. Crick

---

* The term "nucleic acid" is something of a misnomer. It is now known that both DNA and RNA are present as well in prokaryotic cells, which have no nucleus.

counted himself cautiously among this group. By the time he and James Watson met, though, Watson was outright convinced. A blunt American with a crew cut that stood out like a sore thumb at the Cavendish, Watson had been sure of the primacy of DNA since his undergraduate days at the University of Chicago. He was angered by the reluctance of more established scientists to recognize DNA's importance. Later, in *The Double Helix*, he railed against the resistance he had faced from "cantankerous fools who unfailingly backed the wrong horses," adding for good measure that "a goodly number of scientists were not only narrow-minded and dull, but also just stupid."

After Watson arrived in 1950, he and Crick were drawn together by a crystallographer named Maurice Wilkins, who had produced some of the first X-ray diffraction images of DNA. Wilkins soon recruited both men to help him make sense of his raw data.

At the time, prominent labs around the world were racing to discover the structures of proteins. Protein research made up the bulk of the crystallography work being done at the Cavendish. As for solving the structure of DNA, the most serious competition Watson and Crick faced was from a team led by Linus Pauling at Caltech, but Pauling was handicapped by his lack of access to the state-of-the-art X-ray data being produced in Cambridge. By 1953, Watson and Crick were accessing increasingly detailed data provided by a researcher who had taken over the lead role in much of the Cavendish's DNA work: Rosalind Franklin, a chemist and the niece of former home secretary Herbert Samuel. Drawing upon the raw data provided by Franklin's increasingly sophisticated crystallographic work, Crick and Watson were finally able to deduce the structure of DNA.*

---

* The use of Franklin's crystallography has become one of the more enduring controversies over ethics in science. Watson and Crick certainly drew upon Franklin's data without her consent, though it is doubtful that she would have objected. Franklin and Crick grew to be good friends in the years following the discovery. When Franklin died of ovarian cancer in 1958, possibly caused by radiation exposure from her X-ray work, she passed her final weeks at Crick's home. In 1962, Crick, Watson, and Wilkins shared the Nobel Prize in Physiology or Medicine for their discovery of the structure of DNA. Franklin was not under consideration, since the Nobel cannot be issued posthumously, and for many years, her vital contributions to the understanding of DNA were overlooked. The perception of her as victim of an overwhelmingly patriarchal scientific establishment was reinforced by Watson's overtly sexist portrayal of her in *The Double Helix*. Watson called Franklin "the product of an unsatisfied mother who unduly stressed the desirability of professional careers that could save bright girls from

It became one of the most recognizable and beautiful images in science: the twisting double helix of DNA, two long strands of nucleotides wrapped around each other like a tiny caduceus. It was a structure that might have been envisioned by Salvador Dali. Most critically, it had all the attributes that scientists expected to find in the carrier of genes. As early as 1927, the Soviet scientist Nikolai Koltsov had proposed that genes were passed along by a hereditary molecule consisting of "two mirror strands that would replicate."[*] In 1934, Haldane had guessed that genes copied themselves by means of complementary templates. By the time Crick and Watson set out to find proof that DNA was indeed the carrier of genetic information, they knew they were looking for complementary strands that could serve as a template. The structure of the DNA double helix fit like a glove.

In May of 1953, three complementary papers announcing the discovery of the double-helix structure of DNA appeared in *Nature*—one authored by Crick and Watson, one by Maurice Wilkins, and a final paper by Franklin. The structure, Watson and Crick wrote, "suggests a possible copying mechanism for DNA." The articles appeared just a few weeks before Miller published the results of his own experiment in *Science*. Yet in contrast to the barrage of stories that trumpeted the discovery of the generation of amino acids from inorganic elements, the discovery of DNA's structure received almost no notice in the popular press. A short article entitled "Form of 'Life Unit' in Cell Is Scanned" had been slated to run in the *New York Times*, but it was pulled at the last minute, presumably because the editors thought it insignificant.[†]

Despite having just made one of the most important discoveries of the century, Crick was put back to work on the structure of hemoglobin. The idea that proteins still played a significant role in inheritance did not die quickly. Many scientists persisted in the belief that DNA and proteins had

---

marriages to dull men. . . . The thought could not be avoided that the best home for a feminist was in another person's lab." The line is typical of the way Watson described her throughout the book, and a good example of the obstacles women have faced in the laboratory.

[*] Koltsov's theories on genetics led to his denunciation by Trofim Lysenko. In 1940, Koltsov was fatally poisoned by the Soviet secret police. His wife committed suicide on the same day.

[†] History is littered with cases of groundbreaking discoveries being initially overlooked by the lay press and general public. Albert Einstein proposed the theory of relativity in 1905, yet his name did not appear in a newspaper until 1917.

An iconic 1953 image of Watson and Crick with a DNA double helix model.

a kind of symbiotic relationship in controlling the flow of genetic information, that while DNA shared information with proteins, proteins also shared information with DNA, making them cocarriers of genetic inheritance. Crick gradually won support for what he had at first controversially called the "central dogma" of biology: that genetic information can be passed from nucleic acid to protein, but not vice versa. Acceptance of DNA as the sole carrier of genetic information would come only after Crick spent the next thirteen years deciphering the intricate language that organisms have used to speak to each other for billions of years, the genetic code.

THE GENETIC CODE IS the oldest language we know of. It is as old, or at least nearly as old, as biology itself. For billions of years, it has been "spoken" by every cell of every living thing. It has only four letters, each recorded by the presence of a specific chemical. It is typically translated as A, C, G, and T, the letters corresponding, respectively, to the chemicals adenine, cytosine, guanine, and thymine, which are the bases of nucleotides, all arranged in long strings composed of three-letter words.

It should have come as no surprise that the code's eventual unraveling began in Great Britain, the nation where Alan Turing and his Bletchley Park colleagues had turned their skills to breaking German ciphers during the war and constructed one of the world's first computers. With the help of several scientists, including the Russian-émigré physicist George Gamow, most noted for his advancement of the Big Bang theory, Crick and colleagues managed to crack the basic underlying structure of biology's genetic language. By 1966, three years after Crick received a Nobel Prize for his role in establishing the structure of DNA, the code had been deciphered in its entirety, showing how each three-letter sequence, known as a codon, translated into corresponding positions in proteins. With this discovery, human beings could read the cellular language of living things.

As Crick worked on breaking the code, he also confronted the question of *how* exactly DNA communicates with proteins. A language that could not be understood was useless on its own. DNA had to be able to direct proteins to facilitate the sequential arrangement of amino acids. There had to be some kind of messenger between the two. Since the 1940s, some scientists had suspected that large molecules of a nucleic acid called ribonucleic acid, or RNA, played a role in the creation of proteins within cells. By 1958, Crick and others had largely worked out RNA's role in passing genetic information from DNA to proteins. Crick also noticed that RNA played a versatile role in the cell, that it resembled in some ways both DNA and proteins, the agents of replication and metabolism. Though it carried genetic information, he wrote, some RNA must have once been capable of doing "the job of a protein." He even speculated that the first living organism might have "consisted entirely of RNA." His remark would eventually come to be seen as prophetic by many in the field of origin-of-life research.

As the centrality of nucleic acids in genetics began to be established in biology, it started to transform the way scientists looked at the origin of life. If, as it seemed, a particular subcomponent of a cell preceded the others, then either metabolism or genetics had to have developed first. One school of thought, which would come to be called "metabolism first," saw the protein or something like it as the earliest key component of life. In contrast, other scientists, including Stanley Miller, thought that work on proteins was barking up the wrong tree, and that the development of DNA and of genetic machinery was the likely first step. Once replicating, mutable molecules existed, all else would follow through evolution. A protein bereft of a gene, they thought, could lead nowhere.

Sidney Fox always remained firmly in the metabolism-first camp. As most scientists in the field began to tilt heavily to replication or some combination of the two, he started to complain bitterly about what he called the "nucleic acid monopoly." But the biggest problem Fox faced wasn't so much his insistence on the protein-first model for the earliest life. Rather, it was his dogged insistence that he had largely solved the problem of abiogenesis through his experiments on proteinoid microspheres. During the 1970s, Fox became fixated on the existence of differential electric charge that he found on the membranes of the microspheres, which to his mind was not unlike that which exists in living cells. As late as 1988, in his book *The Emergence of Life*, Fox went so far as to claim that these microspheres showed signs of a "rudimentary consciousness."

The claim was met with incredulity and, in some cases, even ridicule in the origin-of-life circles in which Fox had once been so prominent. But he never lost the high esteem of administrators at NASA, and he continued to receive generous funding into the twilight of his career. And although he was an atheist, Fox even managed to secure an official position as an occasional adviser to Pope John Paul II on the subject of the origin of life. By the time of his death in 1998, Fox was mostly ignored by his origin-of-life colleagues, and his unique ability to promote his own work among the organs of power stirred up more than a little jealous resentment.

The true legacy of Sidney Fox's work is more mixed. Fox's skills as an institution builder were instrumental in turning the study of the origin of life into a mainstream academic discipline. When many scientists were leaning toward the idea that the origin of life was a random, unique event, Fox stayed true to Oparin's vision of life's beginnings being part of an inevitable evolutionary progression. And though few scientists still view proteinoid microspheres as significant, the idea of the importance of some type of preprotein, a polymer of amino acids, has never been fully dismissed and would form the basis of many later theories as to how the first organism may have come into being.

LIKE SIDNEY FOX, Francis Crick would, in his later years, face many strains on his reputation. Some stemmed from his outspoken stance on controversial subjects; others, to his penchant for broad, daring hypothesizing. When correct, these ideas reinforced his reputation for genius. When wrong, they could make him appear a bit of a crackpot. He embraced the spirit of the late 1960s, wearing sideburns and colorful shirts and experimenting with LSD. He lent his name to a campaign to legalize marijuana in Britain. He also made several ill-advised comments in support of euthanasia and eugenics, which he came to regret.

Religion was another controversial topic that Crick did not shy away from. After his work on elaborating the nature of genetics, he had been named a founding fellow of Churchill College. It was a prestigious appointment. Named in honor of Winston Churchill, the college was meant to become a kind of British counterpart to American scientific universities like Caltech and MIT. But Crick soon resigned in protest of the building of an exclusively Christian chapel rather than, as Crick preferred, a nondenominational meditation room that could be used by people of all faiths. Crick was not fond of religion in general or Christianity in particular, which, he once joked, "may be OK between consenting adults in private but should not be taught to young children."

Crick moved on to the Salk Institute in California, named for the discoverer of the polio vaccine. There Crick turned his focus more thoroughly to the question of the origin of life. Joined by an old friend from Cambridge, Leslie Orgel, a leading authority in the origin-of-life field, Crick began to contemplate an early stage in the history of life in which amino acids were ordered into primitive proteins using a simple code that might have then evolved into the genetic code used in all modern organisms. He became fixated on the reasons why alternative codes had not arisen, spawning competing lines of descent.

Frustrated with his inability to make headway into the problem of the origin of life, Crick began exploring the idea that life might have originated elsewhere in the universe. In a 1973 article entitled "Directed Panspermia" in the planetary science journal *Icarus*, he and Orgel presented a theory that the Earth had been deliberately seeded with bacterium-like life-forms by an intelligent species from another solar system. Orgel treated the subject almost as a joke. Crick was not entirely unserious, though he knew it was wild speculation. They had based much of their argument on the puzzling abundance of molybdenum in cells. Since molybdenum is extremely rare in the Earth's crust, maybe, they argued, our ancestors emerged on a molybdenum-rich planet. Most scientists were quick to point out that pushing the problem of the origin of life off Earth did little to solve the problem, and quite a bit to complicate it.

But the idea that life might have arisen elsewhere in the universe was not quite dead, and Crick's hypothesis would seem, for a brief spell at least, almost prescient. Two decades after Crick's musing, a rock was found in the Antarctic that would reignite the idea that life might indeed not have originated on this planet, or at least that it was not exclusive to the Earth.

LIFE EVERYWHERE

We are made of star-stuff. We are a way for the cosmos to know itself.

—CARL SAGAN, *Cosmos: A Personal Voyage*, 1980

SHORTLY AFTER MOST of the planets in the Milky Way had formed some four and a half billion years ago, a volcano erupted on Mars, spewing molten lava onto the surface of the planet. As the lava cooled, it hardened into solid rock. For the next half-billion years, the rock lay relatively undisturbed until, one day, an asteroid came crashing onto the planet's surface. The impact was so powerful that the heat from the blast violently compacted the rock, melting away portions of it and creating a series of tiny cracks. The impact also tore the rock away from its resting place below the surface of the planet, and bounced it above the Martian surface.

Four billion years later, another asteroid slammed into Mars. This one struck with such fury that the rock was hurled skyward, through the planet's atmosphere and deep into space. Finally, slowed by the gravitational pull of the sun, and nudged by that of Jupiter, the rock settled into an orbit not unlike that of the planet that had once been its home. It circled the sun for sixteen million years, until one day, just as humans were starting to form permanent settlements along the Euphrates River, its orbit brought it into Earth's gravitational field. The tiny rock, by then no larger than a softball, hurtled down through our atmosphere and embedded itself deep in an ice field in Antarctica.

Another thirteen thousand years passed. Slowly the rock was pushed to the surface by the force of the ice field butting up against a nearby mountain range, like a splinter being forced from a finger. It came to rest in the Allan Hills region of Antarctica, at the base of the Transantarctic Mountains, one of the largest and least explored mountain ranges in the world.

On a relatively mild December day in 1984, a team of meteorite hunters from NASA's Johnson Space Center in Houston began combing the area around Allan Hills known as the Far Western Icefield. Since the early 1970s, NASA had sent dozens of such missions to the Antarctic, which had long been recognized as an ideal place to find meteorites. The extreme environment meant that it was relatively sterile compared to almost anywhere else on the Earth's surface, minimizing the dangers of contamination. And the vast flat sheets of pristine white ice that blanketed most of the Antarctic meant that meteorites were easy to spot.

The person who spotted the Mars rock was a novice meteorite hunter named Roberta Score, on her first such expedition for NASA. It was the middle of summer in the Southern Hemisphere, clear and warm for the coldest point on Earth, almost above freezing. She spotted it in the distance, blue in the sunlight and barely bigger than a softball, likely unnoticeable in any other place on Earth. Score retrieved it and gave it a name: Allan Hills 84001, or ALH84001 for short. Score also took a few notes. She noticed that the rock was "covered with dull fusion crust. . . . Areas not covered by the fusion crust have a greenish-gray color and a blocky texture." Years later, when the rock attracted serious scientific scrutiny, the blocks that Score had observed would become the center of what, for a time at least, seemed like one of the most important discoveries in the history of science.

When the expedition returned, ALH84001 was shipped back to Johnson Space Center. It was placed in the containment facility originally built to contain the lunar samples brought back by *Apollo 11*, where it sat along with the agency's rapidly growing collection of meteorites. Initially, nobody at Johnson suspected that ALH84001 was anything more than a typical meteorite composed of leftover debris from the formation of the solar system, something that had come loose from an asteroid. A chip of it was parceled out for display to the Smithsonian National Museum of Natural History, where it sat, rather unremarkably, for the next five years.

In 1990, a young Smithsonian curator charged with doing some more detailed tests on the composition of the little meteor fragment found that it abounded with carbonate minerals. Although they can be produced through nonbiological means, carbonates on Earth are almost always found in areas that have been exposed to water. This was the first real clue that ALH84001 might be anything but a run-of-the-mill meteorite.

By 1993, mineralogy, isotopic composition, and analyses of trace gases trapped inside the rock had established that ALH84001 was indeed a *Martian* meteorite. It was certainly not the only Martian rock to have fallen onto the Earth. But the technology used to identify extraterrestrial rocks was relatively new, and out of the thousands of meteorites that had been studied by that time, only nine had been conclusively identified as Martian.

ALH84001 began to draw serious attention from labs and research institutions around the world. Pieces were parceled out for study in the United States and abroad. Scientists in Germany were the first to estimate the age of the rock using radiometric dating. They judged it to have formed four and a half billion years ago. This assessment did not quite establish it as the oldest known rock in the universe; another Martian meteor was found to be a tad older. But because of a margin of error that ran in the tens of millions of years, it very well might have been. In any event, over the next several years, ALH84001 became, in the words of one observer, "the most studied 2 kilograms of rock in history."

The biggest piece was reserved for analysis at NASA, where it fell under the responsibility of geologist David McKay. McKay was chief scientist for astrobiology at Johnson Space Center, and an old hand among astrobiology scientists. He had been a doctoral student at Rice University and had stood in the auditorium when John F. Kennedy delivered what would come to be known as "the moon speech" in 1962.* About a decade later, McKay led one of the research groups established to study the lunar samples from

---

* "But why, some say, the moon? Why choose this as our goal? And they may well ask why climb the highest mountain? Why, 35 years ago, fly the Atlantic? Why does Rice play Texas? We choose to go to the moon. We choose to go to the moon in this decade and do the other things, not because they are easy, but because they are hard, because that goal will serve to organize and measure the best of our energies and skills, because that challenge is one that we are willing to accept, one we are unwilling to postpone, and one which we intend to win."

Apollo 11. He would also later play an important role in NASA's *Mariner* and *Viking* missions to Mars.

During the *Mariner* and *Viking* missions, McKay helped spot signs that rivers and lakes had once existed on Mars. The possible presence of water was enticing for McKay and other exobiologists looking for traces of life. Water is essential for life—at least life as we know it. It makes up 80–90 percent of almost all living creatures, and it would be hard for any scientist to conceive of a living thing in its absence. But the *Mariner* and *Viking* missions also found the surface of Mars to be a hostile place, at least from the perspective of Earth life. For billions of years, the atmosphere of the planet had been stripped by the constant bombardment of asteroids, like the one that hurled ALH84001 into space, and the incessant solar wind. The atmospheric pressure on Mars was simply too low to support standing liquid water on its surface. If water had ever existed there, it had long since disappeared, either by evaporation or by sinking slowly into Mars's subsurface.

But ALH84001 was a very old rock, formed at a time when Mars was a very young and very different planet. The carbonate mineral deposits piqued McKay's interest. On Earth, carbonate minerals form almost exclusively in the presence of water. To McKay, that likely meant that sometime early in the history of Mars, water had seeped into the rock. And where there was water, there could very well have been life.

McKay was intrigued by the fact that the carbonate minerals were concentrated on the greenish blotches that Roberta Score had noted when the rock was first found. They resembled the kinds of traces left behind by the *Cryptozoon* found by Charles Doolittle Walcott in the Grand Canyon a century earlier. Soon, NASA scientists discovered something that intrigued them even more: tiny crystals of magnetite. These were embedded in cracks in the rock and concentrated around the markings first observed by Roberta Score. The crystals bore a striking resemblance to those left behind by magnetotactic bacteria that teem in the Earth's oceans.

Humans started using the Earth's magnetic field as a means of navigation and orientation only about a thousand years ago, but a variety of organisms in the natural world have been doing it for millions of years.

There is a great deal of evidence that birds, bats, and bacteria use magnetized iron minerals such as magnetite—or "lodestone," from the Old English for "leading stone"—to orient themselves according to the Earth's magnetic field. The tiny crystals seen in bacteria are called magnetosomes, and they are present in diverse types of microbes, suggesting that these internal compasses are very ancient in origin. Fossil remains of these structures have been found in terrestrial rocks believed to be almost two billion years old.

On Earth, the little grains of magnetite would usually be considered biomarkers, indicators of living organisms. McKay's team decided to send a sample of ALH84001 to Stanford University, where a chemist named Richard Zare had invented a laser mass spectrometer capable of identifying chemical compounds without the invasive and damaging steps necessary in traditional analysis. Zare came back to them with fascinating news: the rocks were filled with organic compounds called polycyclic aromatic hydrocarbons, or PAHs. Although they can be by-products of fossil fuel combustion, PAHs are also often associated with the decay of ancient microorganisms. In private at least, members of McKay's team began to speak about the possibility that they had in their hands proof that life had once existed on a planet other than Earth, what could turn out to be one of the most important discoveries in history.

To make such a startling claim, the NASA scientists needed to identify an actual fossil of a Martian bacterium on ALH84001. But a year of intensive analysis had yielded nothing that looked like any fossil that had ever been found on Earth, so McKay decided to do something nobody else had thought of: he decided to look for something *smaller*.

In January 1996, McKay gained access to one of the most powerful electron microscopes in the world, used by NASA engineers to search for microscopic flaws in hardware on the space shuttle. ALH84001 became one of the first rocks ever examined with the high-power instrument. Over the coming months, McKay was able to make out tiny structures that resembled terrestrial bacteria. But these were *extremely* tiny, the smallest just a hundred-thousandth of a millimeter long. Fifty of the largest, placed side by side, would fit comfortably in a human blood cell. The most inter-

esting of the bunch was intriguingly wormlike. This one would later form
the basis of the seminal photo accompanying their discovery, and it would
be seen around the world.

By early 1996, NASA officials were confident they could make a case
that ALH84001 did, in fact, contain compelling proof that life had existed
at some point in Mars's distant past. McKay and NASA officials planned
to hold a press conference announcing their findings, timed to coincide
with the publication of a paper detailing the discovery that had been submitted to *Science*. Their plans were almost stymied by a high-priced call
girl named Sherry Rowlands. Rowlands had learned of the discovery from
Clinton political adviser Dick Morris, who had been privy to White House
briefings about it. She tried to sell the story to the tabloid *Star*, but the
reporter viewed the story as too fantastic and Rowlands too unreliable.

Rowlands was not the only person Morris had shared the secret with.
Eventually, the story was leaked, and newspapers around the world hailed
the discovery of extraterrestrial life. President Bill Clinton hastily called a
press conference to herald the discovery. "Today Rock 84001 speaks to us

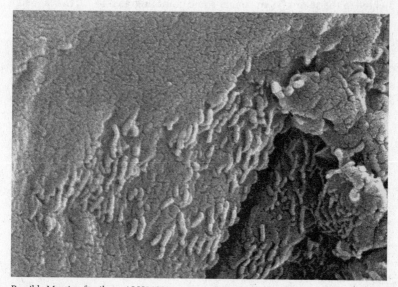

Possible Martian fossils on ALH84001.

across all those billions of years and millions of miles," he announced. "It speaks of the possibility of life. If this discovery is confirmed, it will surely be one of the most stunning insights into our universe that science has ever uncovered."

Yet the final judgment about the significance of ALH84001 by the scientific community was still to come. Over the next few years, a chorus of naysayers would emerge. Many would come to see the episode as reminiscent of the hoopla surrounding another meteorite that had sparked excitement a century and a half earlier.

THE NOTION OF a universe teeming with life is certainly nothing new. As far back as the fifth century BC, the Greek philosopher Anaxagoras had put forth the idea that the seeds of life were scattered about everywhere throughout the cosmos. He called this concept *panspermia*, "life everywhere," and it was later popularized in the poetry of Lucretius. Eventually, the term came to refer to the belief that life had initially developed elsewhere in the solar system, and at some point migrated to Earth via meteorites or as space-borne spores.

The theory of panspermia became particularly widespread in the nineteenth century, when it was championed by two of the most respected scientists of the age: the Swedish chemist Svante Arrhenius, one of the first directors of the Nobel Institute; and Lord Kelvin, whose low estimates of the age of the Earth had stymied Charles Darwin. Kelvin was a particularly enthusiastic supporter of the theory and had made the case that such life was responsible for initially seeding life on the Earth. "We must regard it as probable in the highest degree," he once wrote, "that there are countless seed-bearing meteoric stones moving about through space."

As was the case for many who shared his belief in panspermia, Kelvin's interest had been piqued by one of the most famous meteorite controversies in history. On a spring evening in 1849, a bright fireball was observed streaking over the skies of southwestern France, accompanied by a sonic boom. When it impacted the Earth's atmosphere, it smashed into at least twenty pieces near the small village of Orgueil. One by one, the meteorite fragments were recovered by the villagers, who had watched the whole episode in amazement. A little more than 25 pounds of rock was collected, soft enough that it could be cut with a knife and used to write like a graphite pencil. Soon a parade of top French chemists had taken turns examining the rock, including Marcellin Berthelot, whose work synthesizing organic compounds had made him one of the best-known chemists in the world. Berthelot and many other scientists concluded that the meteorite fragments contained materials of biological origin, and thus that somewhere beyond the Earth, life must also exist.

Eventually, Louis Pasteur was called on to give a definitive opinion as to whether the Orgueil meteorite indeed contained signs of life. It was just five years after Pasteur had convincingly disproved the notion of spontaneous generation, and he was already well on the way to his status as France's experimenter par excellence. Pasteur suspected that contamination on Earth might have been responsible for the startling results seen by Berthelot and others. He decided to build a special drill that would enable him to see whether the organic compounds found in previous studies were present only in the outer layers, and could thus be chalked up to terrestrial contamination, or were also present on the inside, which would presumably show that the substances were indigenous to the meteorite. Pasteur could find no trace of microbes inside the meteorite, and he concluded that there was no evidence for biology in the samples. The clamor of interest that the Orgueil meteorite had excited soon subsided.

The Orgueil meteorite spent the next century as nothing more than a museum curiosity. Then, a pair of American scientists, Bartholomew Nagy and George Claus, decided to take another look. They were surprised to discover small, cell-shaped structures that they interpreted as fossilized alien microbes. In 1962, they reported their findings in *Nature*. The discovery was met by considerable skepticism from almost every credible scientist. The Orgueil meteorite had by then spent nearly a century on Earth in completely nonsterile conditions, sloppily curated in dusty museum drawers. Any clues pointing to the existence of life on the rock had long since been compromised. Nevertheless, a lively debate carried on for two years in the scientific literature.

In 1963, a symposium discussing claims for evidence of life in the Orgueil meteorite was convened at the New York Academy of Sciences. The meeting was presided over by none other than Harold Urey, by then the grand old man of exobiology, and attended by various luminaries in the study of micropaleontology, meteoritics, and the origin of life, including Sidney Fox. As the meeting progressed, some of the objects were debunked as nothing more than common ragweed pollen grains that ubiquitously waft through air. Other particles, however, defied explanation. They were unlike anything found on Earth. But were they truly evidence of extraterrestrial life?

The debate eventually turned to the question of size, the same point of contention that was later resurrected when scientists debated the evidence for Martian life in ALH84001. Some at the meeting were convinced the odd microstructures looked like things that had been alive. Naysayers in attendance said they were no more alive than Sidney Fox's microspheres, and were perhaps derived by the same processes. Urey, ever the open-minded skeptic, concluded the meeting by declaring himself unsure about Orgueil but calling for greater efforts to search for signs of life in meteorites: "The study of carbonaceous meteorites for life-like forms is not an unreasonable pursuit, particularly when one considers that the United States plans to spend 25 billion dollars to put a man on the moon."

In the 1960s and 1970s, interest in panspermia was revived first by the paper in *Icarus* written by Francis Crick and Leslie Orgel, and later by the astronomers Fred Hoyle and Chandra Wickramasinghe. Both ideas came to be seen as borderline absurd in the eyes of their scientific peers.

Crick and Orgel had never really treated their model as anything more than fanciful speculation. Hoyle and Wickramasinghe, however, were deadly serious. They proposed that viruses were constantly being delivered to Earth on meteorites. Such viruses, they said, could have been responsible for the flu pandemic that killed between 50 million and 100 million people in 1918. Certain outbreaks of mad cow disease, polio, SARS, and even AIDS might also have originated off-world.

Hoyle was best known as the scientist who, during a 1949 interview on the BBC, had coined the term "Big Bang" to describe what would become the dominant theory of the origin of the universe. Yet by the time of his work on panspermia, he had become nearly as famous for being one of the last holdouts against the Big Bang theory, even though in the years since he had given it a name, overwhelming observational evidence had made the theory a cornerstone of modern cosmology. Nor was Hoyle's cause helped by the fact that he was also a popular science fiction writer in his spare time. Some called his "viruses from space" idea nothing more than an extension of the plot of his 1957 novel *The Black Cloud*. Given the abundance of competing and far more plausible hypotheses about the origin of viruses, Hoyle and Wickramasinghe's ideas were largely ignored.

A FEW DAYS AFTER NASA's ALH84001 announcement found its way into newspapers in 1996, the evolutionary biologist Stephen Jay Gould penned an editorial for the *New York Times* entitled "Life on Mars? So What?" Gould argued that the discovery of life on Mars shouldn't surprise anyone. He had his doubts about specific claims regarding ALH84001, but the crux of his argument was that life on Mars was no more improbable than life on Earth, which had happened "almost as soon as environmental conditions permitted. . . . The origin of life may be a virtually automatic consequence of carbon chemistry and the physics of self-organizing systems, given favorable conditions and the requisite inorganic constituents."

Were ALH84001 to be seen one day as compelling evidence for ancient life on Mars, two conclusions could be drawn. The first is that the origin of life is relatively easy, having happened independently on two planets in our solar system in a relatively short time during the solar system's early history. The second is that the sudden appearance of life, though still possibly a somewhat improbable event, first occurred on *either* Mars or Earth and was then transported in rocks, blasted off one or the other planet's surface by asteroid impacts, to seed the other.*

Regardless of whether life exists on other planets, scientists are com-

---

* The orbits and the relative strengths of the two planets' gravitational fields suggest that transport from Mars to Earth is much easier than the other way around, leading to the rather startling conclusion that we are all, in fact, Martians. Such an inference cannot be made lightly, but there is experimental evidence that modern terrestrial microbes could at least survive such a trip.

pletely convinced that the universe is rife with organic material. Outlandish theories like those of Hoyle and Wickramasinghe and the controversial nature of claims like those made about past meteorites have tended to obscure the fact that organic molecules are indeed present in outer space—and present in vast quantities. The possibility that the first organic molecules on Earth originated in space, even if it seems outlandish to most people, has become very real to the majority of scientists studying the origin of life.

In some ways, the very word "space" is misleading. The vast expanses of space are not empty, but filled with cosmic clouds of gas and dust. The collapse of some types of cosmic clouds is thought to give rise to the formation of solar systems throughout the universe. We now know that the clouds themselves are filled with organic molecules, some of which likely make their way to planetary surfaces. Before turning to ALH84001, David McKay had been part of the growing effort at NASA to find ways of studying the chemical makeup of cosmic dust.

NASA space missions like *Apollo* and *Viking* have focused on searching for organic compounds such as amino acids on other worlds, but some of the sturdiest proof of a universe abundant in such complicated organics may be meteorite samples already in NASA's possession. Compelling evidence was provided by a meteorite that fell near the town of Murchison, Australia, in September of 1969, just two months after the return of the *Apollo 11* astronauts. It proved to be one of the biggest meteorite finds in history, yielding several hundred pieces that varied in size from a few ounces to 113 pounds. One fragment punctured the roof of a farmer's hay shed.

By late 1969, NASA was well prepared to study the Murchison samples and ensure a minimum amount of terrestrial contamination, eliminating the problems that had plagued meteorite finds ever since Orgueil. The samples turned out to be filled with compounds crucial to biochemistry. To date, ninety-two distinct amino acids have been identified in the meteorite, only nineteen of which naturally occur on Earth. These represent the strongest pieces of evidence that the organic building blocks of life could have come from outer space. The failure of the *Apollo 11* mission to find significant amounts of organic compounds on the moon and the wildly speculative theories of scientists like Fred Hoyle have masked the fact that most serious exobiologists remain quite confident that, at the very least, organic materials are abundant elsewhere in the solar system and that life on other worlds is entirely possible.

LIKE SO MANY earlier claims for life in space over the previous centuries, the sensational news of the ALH84001 findings was eventually met by almost as much skepticism from the scientific community as enthusiasm from the general public. The National Academy of Sciences formed an investigative panel. After two years of review, an article published in *Science* entitled "Requiem for Life on Mars? Support for Microbes Fades" conveyed an emerging scientific consensus that much more had been made of the proof of life in ALH84001 than the facts warranted. Most of the evidence painstakingly compiled by McKay and the other scientists involved—the carbonate minerals, the PAHs, and the magnetite grains—could plausibly be attributed to other, nonbiological origins.

One issue was left unresolved: the microscopic fossils that McKay and others believed they had seen. Even discounting the possibility that the fossil-like structures in ALH84001 were the result of microscopic contamination on Earth, the issue of whether they were fossils at all was something that was bound to be wrapped up in subjectivity. In many ways, it was the same problem faced by those trying to establish the fossil history of ancient microbial life on Earth. Identifying such microscopic fossils has long been often a matter of perception. In the case of ALH84001, the problem was compounded by the fact that the fossils seen by McKay would have been made by microorganisms smaller than any microbe known on Earth. This raised questions about the requisite minimum size required for life.

Those same questions were raised by another discovery, which took place in 1996, just months after the details of ALH84001 were made public. Off the west coast of Australia, an oil-drilling rig had brought sample cores of ancient sandstones to the surface, some as old as 250 million years. They were shipped off to the University of Queensland for study. Four years later, the Australian team led by scientist Philippa Uwins announced the discovery of tiny life-forms that were composed of the standard elements one would expect, including carbon, hydrogen, oxygen, and nitrogen, and that appeared to have cell membranes and stained positively under the microscope for DNA. Uwins called them "nanobes" and suggested they might be a biochemically run-of-the-mill but previously undescribed terrestrial life-form. Yet decades after the discovery of Australian nanobes, nobody could say conclusively whether they were organisms, fragments of organisms, or something else entirely. It is the same dilemma that still surrounds the microfossils that McKay and his team observed on ALH84001. It is hard enough for the scientific community to agree on the authenticity of ancient terrestrial microbial fossils. The burden of proof for extraterrestrial bacteria is naturally, and rightly so, much heavier.

But a growing number of scientists in the field of the origin of life were beginning to realize that fossils are not the only means of looking into the past history of life on the ancient Earth. Soon, one scientist in particular would realize that an imprint of that history was embedded in the genes found in every living cell, and that it could be used to paint some sort of portrait of the earliest living organisms.

ONE PRIMORDIAL FORM

*Nature, displayed in its full extent, presents us with an immense tab-
leau, in which all the order of beings are each represented by a chain
which sustains a continuous series of objects, so close and so similar
that their difference would be difficult to define.*

—GEORGES-LOUIS LECLERC, Comte de Buffon,
Histoire naturelle des oiseaux, 1770

THE YEAR IS 3,500,000,000 BC. The place is a rocky outcropping that juts
out into a shallow, wave-lapped inlet on a landmass that will one day
be called Australia. The seas are bright green and have the sulfurous stench
of rotten eggs. The moon looms large in the sky, twice as large as the moon
does today because it is only half the present distance from the Earth. The
sun, however, is only about three-quarters as bright as it will one day be.
Still, it bombards the Earth's surface with deadly ultraviolet radiation
unchecked by a protective layer of ozone. The atmosphere is filled with
toxic gases, and almost completely devoid of oxygen. That will come much
later, the product of photosynthesis by tiny organisms that will one day
churn away in the primitive oceans.

But the ancestor of those creatures is already here. It lives in the ocean
near the shore, close to a hydrothermal vent that keeps the temperature of
nearby water close to boiling. It is a tiny, single-celled organism, no more
than a lipid membrane that encases an early but functioning genome composed of DNA, as well as proteins and the RNA with which these parts

communicate. Billions of years in the future, scientists will give it a name: LUCA, short for "last universal common ancestor."

Eventually, LUCA will give birth. This will be a virgin birth, which it will accomplish through binary fission. When it is done, LUCA will have divided into two of what will be essentially clones, distinguishable only by the odd genetic mutation. Soon other such clones will appear. They will share their genetic information with each other in a haphazard fashion. Their genetic code will be pooled, their evolution shared. Collectively, they will exist less like a community of organisms and more like a communal organism.

In time, the shallow lagoon will be filled with such organisms, forming small domed masses that peek out from the surface of the water. They will appear to be made of mud. This will be an illusion. Inside, they will be composed of finely laminated layers of silt and biological material. These are microbial mats, not unlike the masses of scum that float to the surface of modern ponds, filled with complex symbiotic communities of micro-organisms interspersed with fine particles of clay and other minerals that have adhered to their cell walls.

Under a microscope, these communities would appear as tiny oval cells juxtaposed with filamentous bacterial forms. They look a lot like stromato-lites, a few versions of which still exist in Australia, as well as on the shores of such far-flung places as the Yucatán, British Columbia, and Turkey. Dynamic and resilient microbial ecosystems, they are capable of thriving in environments hostile to most modern forms of life. In the years to come, these early descendants of LUCA will grow and multiply, carried by winds and currents to distant refuges across the Earth, until one day they will collectively evolve into human beings and every other living thing on the planet today.

EVER SINCE the publication of *On the Origin of Species*, scientists have speculated about what Charles Darwin tantalizingly called "one pri-mordial form." Though based on reasonable evidence, the scenario just depicted is purely hypothetical. There are many such guesses about the

Stromatolites in Australia's Yalgorup National Park.

precise nature of LUCA. The reality is that nobody knows exactly what the environment was like when life arose some four billion years ago. Instead of the hot water of a hydrothermal vent, for instance, LUCA might have lived in a warm little pond not unlike the one once suggested by Charles Darwin. Nor does anyone know what LUCA's internal chemistry was like. It may have possessed unique features that have been lost to its ancestors over the course of several billion years of evolution and natural selection.

Until the late twentieth century, scientists had precious little to go on. The reason modern scientists can draw a reasonable picture of LUCA at all is due in large part to the evolutionary detective work of one man, the biophysicist Carl Woese. One of the most creative, revolutionary, and underappreciated biological thinkers of the twentieth century, Woese upended nearly everything that biologists had once assumed they knew about the earliest organisms on Earth, and laid the groundwork for a new understanding of how those organisms existed and evolved. After his death in 2012, some of Woese's most ardent admirers would invoke comparisons to Einstein and Darwin.

ALTHOUGH CARL WOESE would go on to become one of the world's most important biological thinkers, he never much cared about the life sciences when he was young. Growing up in Depression-era Syracuse, New York, Woese was a painfully shy child who fixated on mathematics. Math offered a respite from the chaotic world around him. It was objective, consistent. In his later years, that shyness would keep him away from most academic meetings or conferences, and it probably contributed a great deal to the relative underappreciation of his work by the general public, as well as to the resistance that his most revolutionary ideas would one day meet in the field of biology.

Woese went on to study at Amherst College, where he did his undergraduate work in mathematics and physics. By the time he arrived at Yale to work toward a doctorate, however, he had turned to biophysics, steered, like so many would-be physicists of his generation, into the relatively new science that had been invigorated by Schrödinger's *What is Life?* Woese's work at Yale revolved around radiation and how it might be used to change the molecular structure of viruses, particularly the one that causes Newcastle disease, which afflicts poultry. After graduating, he spent two years in an unsatisfying and ultimately unsuccessful pursuit of a medical degree before securing a position as a biophysicist at General Electric's primary research laboratory in Schenectady.

Crick and Watson's discovery of the structure of DNA had by then begun to reshape the scientific world's understanding of genetics, and Woese spent the next five years at Schenectady trying to decipher the genetic code. The problem, as he saw it, was one of translation. There were just four "letters" of nucleic acid bases, arranged in three-letter "words." These had to correspond to the twenty amino acids found in proteins, allowing them to be strung together in precise sequences. How this was accomplished was anyone's guess.

Woese turned his attention to the problem of understanding the cellular translators of the genetic code, ribosomes, the large molecular structures composed mainly of RNA that read out specific genetic instructions from DNA into proteins. The little-understood ribosomes had started to interest Woese as early as his Yale studies on bacteria, but at Schenectady he was able to focus on them exclusively.

Though Francis Crick and others would work out the problem of the genetic code before Woese, he would come to understand the code's potential in ways they did not. They treated it is a physics problem to be solved mathematically. Ironically, Woese, who had once rejected biology for mathematics, saw the genetic code as a distinctly biological phenomenon rooted in evolution. He saw that the code might be able to serve as a kind of evolutionary time machine, enabling scientists to peek back through successive generations all the way to the dimmest and most distant period in evolution. Instead of trying to gauge the changes between species by measuring the differences between physical manifestations—searching, as Geoffroy Saint-Hilaire had once done, for the similarities between a human hand and a whale fin—Woese believed that evolutionary links could be more conclusively elaborated by tracing the evolution of the cellular machinery that governed the translation of DNA into protein.

Woese would set himself to this task for the next decade, until he understood the history of life well enough that he could completely reshape what had long been thought to be one of the most unshakable foundations of biology, the tree of life.

The first systematic attempt to classify every living thing in the natural world was made by the Swedish physician Carl Linnaeus, one of the eighteenth century's most influential naturalists. In his groundbreaking 1735 book *Systema naturae*, or *The Natural System*, he sorted organisms into three distinct "kingdoms," consisting of the plants, the animals, and the minerals. By 1758, the tenth edition of *Systema naturae* had grown to include seventy-seven hundred species of plants and forty-four hundred species of animals, all systematically grouped and categorized. These numbers seemed awfully large at the time, but in the two and a half centuries since, estimates of the varieties of species have grown exponentially. By the beginning of the twenty-first century, scientists would suspect that there may be as many as a billion distinct species of bacteria, about three hundred thousand species of plants, and perhaps ten to thirty million animal species, most of which are yet-to-be-discovered insects.

Linnaeus's classification scheme was based on fairly easily observable traits: whether things moved or grew, swam or flew, had fur or backbones. He grouped things by physical similarity. Fossils would eventually provide meaning to the scheme, suggesting an evolutionary pattern that linked the different species and providing a body of evidence upon which Darwin's theory of natural selection would rest.

Darwin was among the first to envision a phylogenetic "tree of life," a family tree that stretched back through every generation to the beginnings of life. He included a sparse and simple drawing of his tree in *On the Origin of the Species*. It consisted of a twig-like branching pattern, with extant species represented by the outermost twig points. According to Darwin's scheme, tracing downward from these points would be like traveling back in time, with junctions in thicker and thicker branches where each twig would have a common ancestor. Humans and chimpanzees, for example, would have diverged from a common ancestor, and if one followed the branch leading to that junction backward, it would eventually join at another juncture with yet another branch leading upward to, say, New World monkeys. One could also travel down the tree through evolutionary time—through the divisions of mammals, the vertebrates, the animals—with the tree gradually narrowing until ending at the single organism that is the root of all existing life. Darwin logically concluded that *all* living things must have been descended from one common ancestor, what he described as "one primordial form."

The notion of universal common ancestry became a central tenet of modern evolutionary theory. It was supported by a number of observations, like chirality (first discovered by Pasteur in his work with crystals), the similarities of cellular structures, and the fact that every organism, from microbes to human beings, uses almost exactly the same genetic code. By the modern era, few reputable scientists would argue against the doctrine of common descent.

In time, Darwin's tree of life was enhanced and reshaped by modern paleontology and radiometric dating. Bones could be dated, and the family lineages of developed species could be more accurately reconstructed. As techniques in microbiology improved, organisms were further divided into single-celled and multicellular organisms, and later into two different categories: organisms with a cell nucleus, called the eukaryotes, and those without, the prokaryotes. Eventually, the living world would be divided into five kingdoms: animals, plants, fungi, single-celled eukaryotes, and prokaryotes. But the pool of evidence for the last two kingdoms was lacking. The fossil record of the most numerous, simplest, and presumably oldest species was glaringly thin, and the place of microbes in the tree of life was shaky. Carl Woese was determined to find a way around that problem.

IN 1969, WOESE WROTE a remarkable letter to Francis Crick. It turned out to be a kind of blueprint for how he would spend the next twenty years, and what he hoped to—and eventually would—accomplish. Woese told Crick that he planned to use DNA to reveal what he called the cell's "internal fossil record," establishing the true relationships between organisms: "By deducing rather ancient ancestor sequences for these genes, one will eventually be in the position of being able to see features of the cell's evolution." He had recognized the potential for molecular biology to use the genetic code to fill in the early gaps in evolution that the fossil record couldn't complete. He planned to do this by sequencing a gene that was common to almost every living thing and then using that gene to trace the history of evolution.

By the early 1960s, the process of sequencing the amino acids of proteins had become routine. Emile Zuckerkandl and Linus Pauling took proteins from modern organisms that could already be placed in the phylogenetic tree and showed that the sequences of the proteins differed depending on how recently the species had diverged according to the fossil record. By measuring the differences between proteins from different organisms—

what they called a "molecular clock"—they could calculate just how long ago organisms had diverged during their evolutionary history.

But not all proteins are common to every species. Woese needed to work with something that was present in *all* known organisms, was copied with a great deal of precision, and mutated slowly enough that it could be easily tracked over billions of years. He settled on genes for a type of ribosomal RNA called 16S, so named after the rate at which it sediments in a centrifuge. The 16S RNA genes were long enough to yield detailed data but short enough that they wouldn't be too difficult to sequence.

By the time Woese began his gene sequencing in earnest, he had moved on from General Electric to the University of Illinois at Urbana-Champaign, recruited for a faculty position by the molecular biologist Sol Spiegelman, who had attended a talk that Woese delivered at the Pasteur Institute in Paris. At Illinois, Woese gathered a small team, the most prominent of which was his postdoctoral assistant, George Fox, who would go on to share credit for the team's most important discovery. Together, they began the complicated process of sequencing the genes for 16S ribosomal RNA, or "16S rRNA" for short.

All the analyses had to be done by hand; automated sequencing was still decades away. Woese and his team adopted a technique that had been developed in 1965 by British biochemist Fred Sanger, one of the rare people to have won two Nobel Prizes. Sanger's process involved dividing the genes for the RNA into small, workable pieces by cutting them with enzymes. The small pieces could then be sequenced, after which the whole molecule would be reassembled and the full genetic sequence revealed. Such work was expensive, and Woese turned to NASA exobiology grants to help fund the research. The work was also painstakingly slow. At the start of the project, sequencing a single 16S rRNA gene could take months. For most scientists, such work would have been tedious in the extreme, but Woese relished it. It was like putting together an enormous jigsaw puzzle.

By the spring of 1976, Woese's team had accumulated complete 16S rRNA sequences from a large variety of bacteria. He and his team then turned their attention to a unique set of microbes known as methanogens.

Unusual microbes to the extreme, methanogens had been named for their strange ability to produce methane as a by-product of metabolizing carbon dioxide and molecular hydrogen for energy. Because of their observable, physical characteristics, scientists had long assumed that methanogens were a form of bacteria, but Woese's genetic data established that they were not bacteria at all. He realized that he and his colleagues had upset the foundation of biological taxonomy. Instead of two ancient lineages—eukaryotes and prokaryotes—there were three, all of which had branched off separately soon after the beginning of life.

Woese called his new family the "archaebacteria" but later shortened the name to simply "archaea," meaning "the ancient ones." He then began redrawing the tree of life. In Woese's tree, everything that would have made up Darwin's tree could be grouped into a mere twig at the end of a single branch. By the time Woese was finished, the tree looked more like a complex, asymmetric snowflake with three distinct branches going in different directions from the base. He called these three main divergences "domains." The discovery came to be known in microbiology circles as the "Woesian revolution."

In 1977, Woese, Fox, and NASA jointly announced the discovery of the archaea in a press release that coincided with the publication of a scientific paper in the *Proceedings of the National Academy of Sciences*. The discovery was greeted with skepticism, incredulity, and, at times, anger. It didn't help that Woese had already developed a reputation with some scientists as an odd recluse who was working on an impossible question. Some saw his data as too fragmented to redraw the map of evolutionary relationships. A few thought he was a little crazy. One of the most influential evolutionary biologists of the twentieth century, the German-American biologist Ernst Mayr, became Woese's most ardent critic, telling the *New York Times* that Woese's work was sheer nonsense. Woese's preferred form of defense against critics was to send repeated letters to the editor, which did little to help his cause. When the first major scientific conference was held to discuss his theories, he was not even invited. He might not have attended even if he had been.

Woese holds a model of RNA in 1961.

By the mid-1980s, Woese's views had nevertheless gained some acceptance. In 1992, he was awarded microbiology's most prestigious award,
the Leeuwenhoek Medal of the Royal Netherlands Academy of Arts and
Sciences, conferred once every ten years. In 1996, Woese, his fellow University of Illinois professor Gary Olsen, and their teams published the first
complete genome structure of an archaean, *Methanococcus jannaschii*.
In an article in *Science*, they concluded that the archaea are more closely
related to humans than to bacteria. "The archaea are related to us, to the
eukaryotes," Woese said in an interview shortly after the discovery. "They
are descendants of the microorganisms that gave rise to the eukaryotic cell
billions of years ago."

Woese's work on *Methanococcus jannaschii* turned the tide of scientific opinion heavily in his favor. The new taxonomy was adopted by most

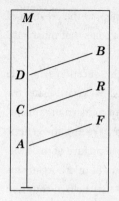

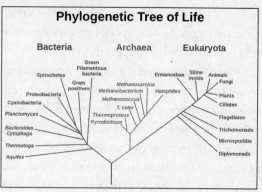

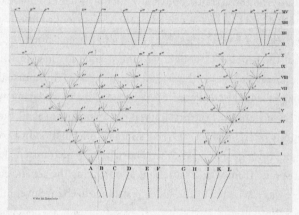

The Tree of Life according to Chambers (top left), Woese (top right), and Darwin (bottom).

scientists. Reflecting on his dramatic vindication, *Science* dubbed Woese "Microbiology's Scarred Revolutionary." The moniker stuck with him for the rest of his life.

THE WOESIAN REVOLUTION had a profound impact on the study of the origin of life. Before Woese, the question of origin was approached mainly from the bottom up. Scientists like Stanley Miller and Sidney Fox had searched for the earliest chemical processes that led, step by step, to the creation of the first living organism, by starting with the earliest steps that must have been necessary for life to appear.

Woese strengthened the case for approaching the question from the opposite direction. The history of life could be traced backward using the evolutionary evidence that still existed in the genes of modern organisms, each of which had evolved from a common ancestor. Tracing the genetic records of different species made it possible to obtain a picture of the earliest ancestor. The characteristics shared by the clusters of microorganisms that made up the bottom of the tree were likely traits they held in common with LUCA. And those traits could show not only what LUCA looked like, but what the world that LUCA was adapted to live in looked like. It was still a hazy picture, but it was far less hazy than it had been before.

One of the problems, though, was the unique manner in which the oldest life-forms had shared their genetic information. This unusual way of sharing genes would lead Woese to make one of his most daring and controversial assertions, that the first life was not an organism at all, but rather a community of organisms that freely swapped information in a collective genetic pool through a process called "horizontal gene transfer." It was an idea that sprang from Woese's attempts at unlocking the genetic code.

Woese was fascinated by the code's resistance to change over billions of years, and through millions of distinct evolutionary branches. With a few very minor exceptions, every modern blue whale, redwood tree, human, and microbe speaks an identical genetic language—remarkable, considering that human languages, over remarkably short timescales, tend to drift into dialects. The genetic code has barely changed at all for billions of years. The reason for its durability remained an enigma as scientists first began to confront the problem in the 1960s.

Woese saw a clue in a discovery that had been made in 1951 by a Seattle physician named Victor Freeman. Freeman discovered that genes from a virus that infects a disease-causing strain of the bacterium *Corynebacterium diphtheriae* could be used to transform a harmless bacterial strain into a virulent one. This finding explained why people infected with diphtheria sometimes don't get sick immediately but often succumb to symptoms later. Perhaps more important, Freeman's observations of *Cory-nebacterium diphtheriae* documented one of the first known examples of horizontal gene transfer, the swapping of genes between distinct organisms. Eventually, many—perhaps even most—microorganisms would be found to use horizontal gene transfer. Woese hypothesized that this mode of genetic exchange explained why the genetic code had not changed in billions of years. Organisms needed a stable, uniform language with which they could genetically "speak" to each other. If their genetic code changed, the beneficial trait of being able to swap genetic information with other organisms would be lost.

The discovery of horizontal gene transfer also solved an evolutionary problem. Darwin's concept of natural selection works best in complex modern organisms that reproduce sexually. Parents have children whose genes are a mixture of the genes of both parents, as well as various accumulated mutations. But most microorganisms produce asexually.* With only one real parent, the offspring are essentially clones, and variation arises only from the mutations caused by environmental insults such as radiation or copying errors. The effects of such mutations are usually unnoticeable, often bad, and sometimes lethal. Every so often, though, they are positive. On rare occasions, they even provide an adaptive advantage for an organism. But relying solely on the accumulation of mutations is a very slow way for evolution to proceed.

The traditional picture painted by biology was one of complex organisms—the higher eukaryotes—rocketing along the evolutionary ladder by sexual reproduction, while lowly prokaryotes were left to plod along with whatever mutations were provided by random chance. Horizontal gene transfer provided an explanation for the rapid evolution of the earliest microorganisms.† By swapping information freely, similar to the way *Corynebacterium diphtheriae* does, early microbes would have been able

---

* There are rare exceptions among higher organisms. Some plants and a few species of insects, fish, lizards, and birds occasionally reproduce through "parthenogenesis," Greek for "virgin birth." Parthenogenesis has even been induced in laboratory settings in the eggs of several mammals, including humans, though the offspring are always unhealthy or nonviable.
† Horizontal gene transfer also seems to be a major reason why bacteria are able to adapt so quickly to antibiotics.

to speed their own evolution by drawing upon a much larger gene pool. Woese called horizontal gene transfer "the tide that lifts all boats."

Eventually, biologists would come to understand that genes are swapped between very distantly related species of bacteria, and that genes can even be scavenged from dead bacteria. When organisms die, their genetic material can linger in the environment for significant periods before decaying. Microorganisms are capable of actually ingesting this discarded DNA and incorporating it into their own genomes. Bits of genetic information are quite literally everywhere in the environment, the Earth being like a giant library from which microbes can borrow.

Woese's last observation about horizontal gene transfer was that it not only complicated the early part of the tree of life but upended it all together. With primitive organisms sharing information so freely, the root of the tree of life was really more like a web, connected not by traditional straight-line patterns of descent but by crisscrossing connections established by horizontal gene transfer. Woese's revisions to the structure of the base of the tree undermined Darwin's idea of a single common ancestor from which all modern organisms spring. In Woese's view, tracing the tree back to a *single* ancestor was impossible. The deepest we can see back into that line is a roiling mess of indiscrete organisms evolving in an interdependent fashion.

PROBABLY THE MOST striking thing about the microorganisms that populated the base of Woese's new tree of life is that many are extremophiles, microorganisms capable of thriving in environments that would be deadly to most modern life. Sometimes popularly known as "superbugs," varieties of extremophiles have been found living at temperatures far below freezing. Some species of a type called acidophiles live off gases dissolved in raw sewage and excrete acid powerful enough to destroy modern sewer systems.

Many of the extremophiles at the base of Woese's tree were hyperthermophiles. Capable of living in intensely hot aquatic environments, hyperthermophiles were first identified in 1965 by the American microbiologist Thomas Brock, who found them living in hot springs in Wyoming's Yellowstone National Park. Since then, seventy separate species have been identified, some in hotter-than-boiling deep-sea hydrothermal vents.*

For decades, discoveries in geology had been hammering away at the Oparin-Haldane "primordial soup" model of the ancient Earth. Geologists had developed increasingly sophisticated methods of deducing what the early atmosphere was composed of by studying basalts, the rocks left behind by volcanic eruptions. While Oparin envisioned an ancient atmosphere filled with ammonia and methane, it seemed more and more likely that the atmosphere was filled instead with nitrogen and carbon dioxide. By the late 1970s, much of Oparin's model appeared suspect, as did the underlying mechanisms that Miller's experiment had suggested. To many, a new model was sorely needed.

Many scientists saw the presence of extremophiles at the base of Woese's tree as clues they could use in reimagining the earliest life-form. These were presumably the closest descendants of LUCA, so it stood to reason that their ability to thrive in extreme environments had implications for the kind of environment LUCA arose in. This speculation led to some radical theories. The most resilient of these came from an unexpected source, a Munich patent attorney named Günter Wächtershäuser, who saw metal sulfide mineral surfaces as ideal places for life's earliest chemistry to get started.

Wächtershäuser was a friend of Carl Woese with whom Woese had shared his growing doubts about Oparin's primordial-soup model. An organic chemist before he had turned to the practice of law, Wächtershäuser put his mind to the origin-of-life problem. At the home of their mutual friend, the philosopher Karl Popper, he shared with Woese his novel concept that life began in deep-sea hydrothermal vents, the first of which had been discovered by the research submarine *Alvin* off the coast of the Galápagos in 1977. Woese was intrigued, and he encouraged Wächtershäuser to work out his model in more detail. Wächtershäuser imagined a series of chemical steps beginning on mineral deposits near

* Because of the high pressures found in deep-sea environments, water there boils at well over 100°C.

hydrothermal vents in the depths of the ocean, where they would have been protected from the hostile environment above. His belief that life began on iron sulfide mineral surfaces gave his model its name: the "iron-sulfur world."

Wächtershäuser's theory gained a strong foothold among adherents of "metabolism first" scenarios, those who saw the evolution of genetic material as something that had happened relatively late in the development of life. He also drew on another implication derived from Woese's redrawn tree of life. Most of the organisms at the base were autotrophs, with unique metabolisms that function by consuming inorganic substances like carbon dioxide and hydrogen sulfide. An autotrophic origin would sidestep the need for abundant organic molecules that the first cells could "feed on," which had been a central tenet of Oparin's model. Finally, Wächtershäuser believed that the most primitive life lacked a cell membrane. This last contention was perhaps the most dubious for more established origin-of-life scientists, most of whom believed some type of chemical barrier was necessary to allow the first living systems to grow more complex.

Other scientists, notably geochemist Mike Russell, would build on Wächtershäuser's theory to try to account for the perceived need for a protective cell-like container. Russell and his colleague Allan Hall, both experts in iron sulfide deposits, liked the idea of mineral-rich hydrothermal environments that Wächtershäuser had proposed, but they speculated on the existence of a gentler kind of hydrothermal vent, a great deal cooler than any of those previously observed. The hypothesis received a huge boost in 2000 with the *Argo*'s discovery of the Lost City, which turned out to be just the kind of hydrothermal field Russell had envisioned. Russell and Hall also came up with a model for the development of a primitive stand-in for a cell membrane. When the alkaline water from these types of hydrothermal vents mixes with the more acidic ocean water, it forms bubbles made of sulfides and other mineral types. These bubbles could have served as primitive membrane-like compartments.

Still, many in the field remained unconvinced. Even Russell's gentler vent was a difficult environment for life to have emerged in. Years after Wächtershäuser first conceived of the iron-sulfur world, only tantalizing glimmers of experimental support had emerged. A large number of scientists believed the key was a molecule that Carl Woese had mused about as far back as 1967. In his first and only book, *The Genetic Code*, he had suggested the possibility that RNA played a much more important, versatile, and ancient role in the cell than many had believed.

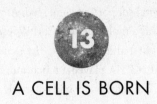

A CELL IS BORN

An honest man, armed with all the knowledge available to us now, could only state that . . . the origin of life appears at the moment to be almost a miracle. . . . But this should not be taken to imply . . . that it could not have started on the earth by a perfectly reasonable sequence of fairly ordinary chemical reactions. The plain fact is that the time available was too long, the many microenvironments on the earth's surface too diverse, the various chemical possibilities too numerous and our own knowledge and imagination too feeble to allow us to be able to unravel exactly how it might or might not have happened such a long time ago, especially as we have no experimental evidence from that era.

—FRANCIS CRICK, *Life Itself,* 1981

IN 1986, AN ARTICLE by the Nobel Prize–winning physicist and biochemist Walter Gilbert appeared in the influential "News and Views" section of the journal *Nature.* Eight years earlier in the same section, Gilbert had hypothesized the existence of sections of genes called introns that were spliced from their RNA carriers during the process of gene translation to proteins, solving a long-standing problem in biology. This time he was suggesting a solution to an even more important mystery, one that had plagued origin-of-life scientists for decades: the chicken-or-egg dilemma of replication or metabolism.

At some point in the half-billion years or so of evolution that took place before LUCA, an even more primitive organism must have existed, what scientists sometimes call FLO, or the "first living organism." Such an organism would have been little more than a simple chemical entity, possibly consisting of a single component of the complicated machinery that makes up all modern cells. But which component? At the time scientists like Stanley Miller and Sidney Fox first pursued the question, the answer had seemed rather straightforward: it had to be protein, since proteins were then still thought by many to be both the principal agents of metabolism and the carriers of genetic information. The work of Crick, Watson, and others in elaborating DNA's role in the cell shifted the focus of many to the principal agent of genetics, which might by itself have been capable of initiating the long chain of evolution. DNA was mere information, though, incapable of accomplishing much on its own. Yet the idea that the first protocell contained DNA *and* protein seemed too complex. One or the other had to have come first.

In his 1986 *Nature* article, Gilbert argued that the solution was that neither had come first. Instead, Gilbert returned to an idea first floated by Carl Woese in 1967. In his book *The Genetic Code*, Woese had speculated that RNA, the intermediary between DNA and proteins, might have once done the jobs of both. This speculation was followed by a pair of papers published simultaneously in 1968 by Francis Crick and Leslie Orgel, both of which postulated that life was initially based on RNA. Orgel's paper, "Evolution of the Genetic Apparatus," would eventually come to be recognized as one of the most elaborate early representations of the idea, although it initially gained very little traction. By 1986, however, scientists had achieved a much deeper understanding of RNA. Gilbert was able to revive the idea because of a groundbreaking discovery independently made during the previous few years by a pair of microbiologists named Thomas Cech and Sidney Altman.

In 1978, Cech, a relatively young assistant professor at the University of Colorado, Boulder, began a series of experiments to isolate the protein that was responsible for the gene splicing that Gilbert had recently discovered, the removal of introns from RNA molecules. Cech imagined that the protein would be rather easy to find. He would simply take cell extracts and purify them until he found the culprit that was causing the splicing. But his research team was stymied. Even with samples they were sure were entirely free of proteins, the RNA still ended up being spliced. Eventually, they were able to prove that the RNA *itself* was responsible for its own splicing.

Not long after Cech began his work, Altman, a researcher at the Medical Research Council Laboratory of Molecular Biology in Cambridge, then led by Sydney Brenner and Francis Crick, began his own study of a strange enzyme called ribonuclease P. Ribonuclease P was unusual because RNA seemed to make up about 80 percent of its mass, which scientists had always discounted as an unimportant anomaly. Altman kept up the work when he moved on to a professorship at Yale, and eventually he was able to conclude that RNA was, in fact, the critical catalytic component of ribonuclease P. RNA, not protein, was responsible for the observed reactions.

Cech and Altman's discoveries revolutionized our understanding of biochemistry. Some proteins catalyze chemical reactions. Some are motors. Still others, embedded in cell membranes, open and close channels that make it possible for the phenomena of consciousness to take place. Eating, digesting, moving, even thinking—all are, at their most basic level, functions of proteins. Proteins control the chemical processes responsible for virtually everything an organism does at the cellular level, and proteins had always been understood to be the sole cellular components carrying out these reactions. But Cech and Altman proved that some RNA molecules—which came to be called ribozymes—could also act as catalysts like proteins do. The discovery earned each a share of the 1986 Nobel Prize in Medicine.

As Gilbert pointed out in his *Nature* article, the discovery of ribozymes had huge implications for the study of the origin of life. Here was a part of the cell that could accomplish *both* replication and metabolism. He argued that at some point very early in the history of evolution, the simple cells that populated the Earth merely contained RNA. "The first stage of evolution proceeds," he wrote, "by RNA molecules performing the catalytic activities necessary to assemble themselves . . . [eventually] using recombination and mutation to explore new functions to adapt to new niches." The theory came to be known by the phrase with which Gilbert had titled his influential article, "The RNA World."

BOTH CECH AND ALTMAN had come to their groundbreaking revelations about RNA while working with samples taken from a microscopic proto-zoan called *Tetrahymena thermophila*. *Tetrahymena* is a remarkable little creature, part of the ciliate family, eukaryotes first observed by Antonie van Leeuwenhoek and characterized by the stringy cilia that spring from them like clumps of hair. *Tetrahymena* comes in seven different "sexes" and is capable of thriving over a wide range of temperatures. But what particularly sets it apart from other single-celled organisms is the wide variety of biological processes it shares with more complex organisms. It has a primitive digestive system, with a mouth-like pore for ingesting food. Remarkably, the tiny organism contains about twenty-five thousand genes, nearly as many as human beings have. The presence of such a wide array of

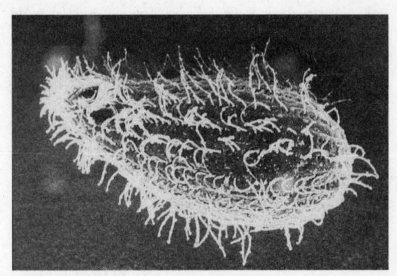

A *Tetrahymena* as seen by a scanning electron micrograph.

genes makes *Tetrahymena* useful for modern biological research, as does the ease with which it can be cultured.

Tetrahymena had already been at the center of an absurdly large number of important biological discoveries, including the identification of the first cytoskeleton-based motor protein (a primitive muscle-like protein) and the existence of lysosomes and peroxisomes, which are like the cell's little wastebins. In fact, the very same sample of *Tetrahymena* used by Cech and Altman was also being used by a completely different group of researchers who would go on to win a Nobel Prize for a completely different discovery. One of those scientists, a young, mild-mannered Canadian named Jack Szostak, would go on to build upon the discovery of ribozymes to become one of the leading origin-of-life scientists of the twenty-first century, and the most famous of the modern scientists who actively study the RNA-world model.

THE SON OF A PILOT in the Royal Canadian Air Force, Szostak was captivated by the *Apollo* missions during his childhood. But he had always been more interested in the experiments the astronauts would conduct on the moon than in space travel itself. Biology had been a particular area of interest. In school, he was a prodigy. He enrolled in McGill, Canada's most prestigious university, when he was just fifteen years old.

After joining Harvard Medical School as a professor of chemistry in 1982, at the age of twenty-seven, Szostak first turned to the subject of DNA repair in the yeast *Saccharomyces cerevisiae*, a model eukaryotic organism that had fascinated countless scientists since Pasteur's time. Soon thereafter, he attended a lecture given by UC Berkeley molecular biologist Elizabeth Blackburn regarding the genetics of *Tetrahymena*. Szostak felt Blackburn's advances could be brought to bear on work he was doing in his own lab that involved a long-standing problem in eukaryotic cell biology: Because the enzymes that copy DNA never quite reach the ends of chromosomes, scientists had always expected that part of the chromosome should remain uncopied at the end of each round of cell division. That this was not always true had long left scientists puzzled.

Along with molecular biologist Carol Greider, Szostak and Blackburn devised a set of experiments to figure out what was happening. By making *Saccharomyces-Tetrahymena* hybrid chromosomes, they were able to show that tiny pieces of DNA called telomeres, Greek for "end part," from the ends of *Tetrahymena* chromosomes could protect *Saccharomyces* chromosomes from being shortened, and vice versa. This finding not only explained the paradox; it had significant implications for understanding cellular aging and cancer. Szostak, Blackburn, and Greider's work was published in 1982. In 2009, nearly three decades later, the three shared a Nobel Prize for the discovery.

AFTER HIS WORK WITH telomeres, Szostak began looking for a new scientific challenge. Very early in his career, he had decided he wanted to work on what he considered the three big questions in science: the origin of the universe, the origin of consciousness, and the origin of life. He realized early on that his math wasn't up to par to tackle the physics associated with the origin of the universe. And though, like Henry Bastian and Francis Crick, he was tremendously enticed by the idea of unraveling the phenomenon of consciousness, he felt the technology wasn't yet there to make much of an impact. But after Cech and Altman's discoveries about the nature of RNA, Szostak saw an opportunity to make important strides in understanding the origin of life.

As early as 1984, Szostak had started to immerse himself in the study of ribozymes, trying to better understand what their roles might have been in the earliest cells. His goal switched to finding something that was held up as a kind of holy grail among those who subscribed to the RNA-world theory: an RNA molecule that could do the work of copying itself. Nothing like it existed, or had ever existed, as far as anyone could prove. But Szostak had new avenues to approach the problem that hadn't been available to many of his predecessors.

Since the Miller-Urey experiment, re-creating the chemical steps that would have led to FLO's initial appearance had proved perplexingly difficult. Though some progress had been made, few scientists had had much success. But Szostak recognized that there might be another way of approaching the problem. By the 1990s, with advances in knowledge about the cell and the development of modern techniques to manipulate the cellular machinery, it became possible for scientists to consider building a cell from scratch. Instead of trying to re-create all the difficult chemical steps necessary for the emergence of the first life, Szostak was simply trying to create it in his laboratory.

THE ERA OF completely synthetic life-forms began in 2002 when a research scientist at a laboratory in Long Island injected the contents of a syringe into a small white mouse. Within minutes, the mouse was dead, its body frozen by paralysis, the telltale sign of the lethal dose of poliovirus it had just received. The poliovirus's deceptively simple RNA-based replication cycle had gone into overdrive, hijacking the mouse's cells into making countless copies of itself, until the host cells, each swollen with ten thousand new viruses, ruptured, releasing them to infect more host cells. But what made this particular poliovirus attack so remarkable was that it had been induced using viral DNA built from scratch in the laboratory, the brainchild of a Stony Brook University virologist named Eckard Wimmer. The poliovirus genome had been deciphered in the summer of 1981, and Wimmer's team had only to download the genetic recipe—a simple string of about seventy-five hundred A's, G's, C's, and U's—off of the Internet.

The technology for synthesizing DNA had been perfected in the previous decades. Building such molecules, while not exactly child's play, was well within the repertoire of the modern molecular biologist by the turn of the millennium. Geneticist Craig Venter and his team started work on synthesizing an entire bacterial genome, a feat vastly more complex than Wimmer's simple virus. Even the smallest and simplest cells consist of hundreds of highly evolved enzymes, as well as the genetic code and all the other trappings used by modern organisms. Though it took twenty-four scientists ten years and forty million dollars, they succeeded, via a complex series of laboratory manipulations, in synthesizing their bacterial genome in 2010. It contained a staggering 1,077,947 base pairs.*

---

* Venter and his team encoded a series of quotations in the synthetic genome as a puzzle. These included a quote from James Joyce's *Portrait of the Artist as a Young Man*, "To live, to err, to fall, to

To make their bacteria, Venter and his team began by adding the synthetic chromosomes they had built to a culture of natural *Mycoplasma* that was subjected to an electric shock. The shock allowed the artificial chromosome to enter the host cell. Then, as the host's cellular machinery went to work on the synthetic genome, daughter cells were produced containing only the artificial chromosome. This chromosome, having been endowed with all of the instructions for making all of the proteins needed to keep the cell running indefinitely, took over, and thus the first completely artificial organism came to life. They named it "Synthia." It contained about eighty times as much genetic information as the poliovirus.

The implications for the future of biotechnology were truly remarkable. The potential applications reached into fields as diverse as synthetic fuel production and medicine. But many scientists were quick to point out that Synthia did little to answer the question of where the information required to build a cell came from in the first place. Venter, like Wimmer before him, essentially copied the blueprint that nature had provided them after four billion years of molecular tinkering. It was an astonishing feat of engineering but proved nothing about *how* life began.

SZOSTAK BEGAN CONTEMPLATING a synthetically engineered cell as far back as the mid-1990s. What set his vision apart from that of men like Venter and Wimmer was that he wanted to understand the *origin* of life, not merely copy the blueprint provided by nature. The key question for him was how the blueprint would arise from scratch. He had some clues to draw upon, provided by a remarkable set of experiments carried out all the way back in the 1960s by Sol Spiegelman, the biochemist who had originally recruited Carl Woese to the University of Illinois.

triumph, to recreate life out of life." Two other quotes included were "See things not as they are, but as they might be," from *American Prometheus*, the story of Robert Oppenheimer and the creation of the atomic bomb, and "What I cannot build, I cannot understand," reportedly the last words left written on the physicist Richard Feynman's blackboard at the time of his death. They also encoded a link to a website where amateur cryptologists could report their having unraveled the puzzle. Venter and his team had a good practical reason to include the quotations: the coded messages were proof of the bacteria's synthetic origin.

Spiegelman and his colleagues performed a series of telling experiments that showed how RNA molecules might behave like organisms and evolve on their own—in a test tube—in a Darwinian fashion. Spiegelman started with a virus known as bacteriophage Qβ—pronounced "cue beta"—which infects the common gut bacterium *Escherichia coli*. They purified Qβ's RNA genome and the protein that copies it, and then mixed these together in a test tube along with the precursors that the protein uses to construct a new Qβ RNA molecule. They let this mixture react for a little while and then transferred a few drops of the solution, now containing various imperfect copies of the original RNA molecule, to a new test tube that contained only the protein and precursors. They did this a total of seventy-four times, each time transferring a few drops from the last test tube to the next. In each exchange, a new population of mutant molecules was transferred, serving as the starting point for the "molecular evolution" that would take place in the next test tube.

The end of this process revealed something remarkable: the starting RNA molecules were about 4,500 nucleotides long, but the ones present in the seventy-fifth test tube were only 218 nucleotides long. There appeared to be a sort of competition in which shorter molecules tended to win. This made perfect sense, since short molecules could be copied faster, and thus could exponentially outcompete longer molecules. In essence, Spiegelman had created a form of natural selection among naked strands of RNA in his test tubes. His colleagues dubbed his creations "Spiegelman's monsters."

In 1975, two scientists in chemist Manfred Eigen's lab performed a surprising experiment that built upon Spiegelman's work. This time they simply mixed the precursors and the protein without the master RNA template. Astonishingly, they found that over time, molecules very similar to Spiegelman's minimal RNA molecule appeared. This result showed that, given the right conditions, a meaningful information-containing molecule, like RNA, *can* spontaneously arise. The results obtained by Spiegelman and his colleagues became a touchstone for origin-of-life scientists in the RNA-world camp. The only problem was that the molecules were not truly self-replicating. They could make copies of themselves only in the presence of the copying protein, itself the complex product of a biological blueprint. Thus, the concept of a *self*-copying RNA molecule became the holy grail of RNA-world research.*

Though various teams had tried to find such a self-replicating RNA molecule, Szostak envisioned taking the concept one step further by building a true RNA organism. Building on an idea he had come up with alongside Italian biochemist Pier Luigi Luisi in 2003, Szostak wanted to work on building an actual RNA-based *cell*, one in which the RNA was housed in a lipid membrane.

According to Szostak's model, the first living thing might have been a naked RNA molecule. But the first living cell—FLO—would have needed a membrane for early evolution. Szostak saw two good reasons for this requirement. The first was that molecules that stay together evolve together. It seems unlikely that a single strand of RNA could do *all* of the various functions needed to propagate a cell. A small family of molecules might have been required, held inside a small bubble that forced them to collaborate. These molecules would copy each other and influence each other's evolution in mutually beneficial ways.

The second, closely related reason was the problem of freeloaders. An RNA molecule that was generally good at copying other RNA molecules in its environment would be beset by parasitic RNA molecules, few of which might actually contribute to the welfare of the pack. Enclosing the copying ribozyme in a cell with only its collaborators—the freeloading viruslike RNA molecules locked safely outside—would make the encapsulated ensemble much more efficient.

In all modern cells, membranes are composed of molecules known as amphiphiles. These molecules are made up of fats, like animal tallow and coconut oil, which form the basis of almost every soap and shampoo. When placed in water, under the right conditions, they have a natural tendency to spontaneously form bubbles ranging in size from the microscopic to the very large, which could have formed the basis of simple cell membranes. There may be good reason to suspect that such lipid bubbles were abundant

---

* There has been some progress in synthesizing such a molecule. In 2011, a British team led by Philip Holliger announced the synthesis of an RNA polymerase, made out of RNA—not protein—that could copy another RNA molecule ninety-five nucleotides long.

on the primitive Earth, even before life had devised ways of making them. Meteorites like Murchison have been found to contain similar molecules, and small bubble-like structures form from the organic "goo" common in such meteorites. This goo could have spontaneously formed cell-like compartments in shallow pools on the Earth.

To realize this vision of FLO, Szostak developed a novel approach harnessing the power of natural selection. Using billions of small random RNA molecules encased in tiny lipid bubbles, he simply lets them compete in test tubes just as Spiegelman's monsters did. The ones that copy themselves well should grow at the expense of those that do not. Szostak hopes that, at some point, a set of these RNA molecules will spontaneously solve the problem of making each other and the membrane itself from simpler precursors that the scientists supply. The result will be a living, evolvable RNA-based organism: an RNA-world re-creation of FLO.

A COUPLE OF BLOCKS from Boston's Charles River, on the seventh floor of a research building owned by Harvard Medical School, a microscope sits in a room barely larger than a walk-in closet. On the door hangs an old photograph of Mahatma Gandhi. Dressed in traditional garb, the Hindu holy man and leader of India's struggle for independence is bent over and peering into a small microscope. Inside, the room is barren except for a single chair and a small table upon which sits an actual microscope. Above it, a picture cut out of a magazine is taped to the wall. It is a picture of a protocell as imagined by Szostak, brightly colored and simple, devoid of all the typical machinery of the cell except for a single strand of RNA. Next to it are four simple words: "A Cell Is Born."

If Jack Szostak and his team are successful in their quest to create a living model of his vision of FLO, it is there, in that little room, that the first self-replicating RNA-based cell will be seen by a human being. It will be a species more primitive than any currently living on Earth, a monumental discovery. It will undoubtedly be greeted by a flood of sensational press coverage. In the eyes of many, the problem of the origin of life will have been largely solved.

Yet history shows that a great enigma will likely remain. It will be the same dilemma embodied by the question once posed by the physicist Enrico Fermi to his old friend Harold Urey: Is this how it *could* have happened, or how it *did* happen?

Carl Sagan was fond of recalling a story from a public panel discussion he had taken part in Chicago in 1960. A member of the audience had asked the panelists when scientists would solve the mystery of the origin of life by actually duplicating the process in a test tube. The first panelist said it would be in about a thousand years. The second said three hundred. Gradually, the numbers got smaller and smaller until one scientist said it had already been done.

The question of the origin of life has always provoked unrealistic expectations in those who look to science to explain the natural world. Similar to the early Christians who waited for a Rapture they were sure was just around the corner, so, too, have believers in science expected that the solution to one of its greatest mysteries was on the verge of being revealed. And so have they waited for hundreds of years.

Depending on one's perspective, the answer to the question of how life began has always been right around the corner—or so incredibly difficult that it may never be definitively answered. Science has no doubt made enormous strides toward understanding the transition from nonlife to life. The mystery has attracted some of the world's greatest minds; no doubt it will continue to do so. It could well be that definitive answers *are* just around the corner. There is likewise a good chance that the question will remain unsolved, at least in our lifetimes. We simply do not know whether the transition from nonlife to life took a week or a month or five hundred million years. A process like this may require such a vast amount of time that it will never be observable in a laboratory.

We do know that science will never stop searching for the answer. Perhaps that search has already yielded something important. Perhaps it has told us something about the nature of science and even of ourselves.

EPILOGUE

Men in their generations are like the leaves in the trees. The wind
blows and one year's leaves are scattered on the ground; but the trees
burst into bud and put on fresh ones when the spring comes round.
In the same way one generation flourishes and another nears its end.

—HOMER, *The Iliad*, c. 1250 BC

I N HIS FAMOUS LECTURE at the Sorbonne, Louis Pasteur made an observation about the nature of science and the role of the scientist. Science, he said, is an impartial arbiter. The true scientist must strip himself of all preconceived ideas. About the subject at hand, spontaneous generation, Pasteur said one could come to no other conclusion through science than to arrive at the necessity of a divine role in the creation of life.

Roughly a century and a half later, the evolutionary biologist Richard Dawkins stood in the laboratory of one of the world's most influential geneticists and delivered a speech that began in almost identical terms. Science, he said, does not take sides; it looks only for objective truth. But here Dawkins, an ardent atheist, diverged from Pasteur, a Catholic and vitalist. For Dawkins, one could draw no other conclusion than to say that the first emergence of life was purely the product of natural laws, completely devoid of the supernatural.

Dawkins was standing in the laboratory of Craig Venter, who had helped map the human genome and had led a team that actually reverse-engineered a living organism. Pasteur delivered his lecture at a time when there was as yet no notion of the existence of nucleic acids and the gene was no more than a concept. The world has changed dramatically since Pasteur's day. Synthetic biology has opened up the inner machinery of living organisms. The engines that drive life can be seen, scrutinized, and manipulated. In our lifetimes, the cost and ease of making a synthetic living thing may become so trivial that it can be something amateur scientists do in their garages. We understand the world and the forces that govern it far better than human beings just a century ago could ever have imagined.

In many ways, our understanding of the origin of life has always been a function of the available tools and technology with which we could make sense of the world. The microscope opened up new worlds to the human beings of the seventeenth century. The printing press of the nineteenth century opened up those worlds to vastly more people. The explosion of technological innovation in the twentieth century—radiometric dating, gene sequencing, exploration of the solar system, to name just a few—has likewise transformed our understanding of biology, and of the origins of biology on this Earth.

But technological progress is not all that has changed since Pasteur's famous Sorbonne lecture. Speaking at Francis Crick's funeral in 2004, Michael Crick said that his father had wanted to be remembered for finally putting to rest the theory of vitalism, the idea that some uncrossable chasm existed between the living and nonliving. Noting that the word "vitalism" was not recognized by Microsoft Word, he said, "Score one for Francis."

The balance of power between science and theology in society has shifted. Nothing symbolizes that shift more than a trip Sidney Fox made to the Vatican in 1964 to advise Pope Paul VI on evolution and the most modern scientific concepts of the origin of life. The pope's predecessor, Pius XII, had already announced that evolution was not incompatible with church teachings, marking a slow but steady return to the spirit of Saint Augustine, who some two millennia earlier had said that the church would not be served by appearing ignorant of the natural world. The march toward an acceptance of scientific explanations for the origin of life seems inevitable. In 1996, Pope John Paul II alluded to the "recognition of evolution as more than a hypothesis." Less than two decades later, Pope Francis warned against thinking of God as "a magician, with a magic wand able to do everything."

In a sense, the world has come full circle. With the notable exception of the United States, religion in most of the developed world has largely ceded to science much of the responsibility for explaining how the physical world functions. Gone are the days when Redi and van Helmont had to tread carefully in the shadow of religious authority, or a man like Robert Chambers was forced to publish anonymously for fear of religious retribution. Even in America, where biblical literalists remain an influential minority, scientists are almost entirely free to follow the pursuit of knowledge wherever it may lead.

TODAY, TENS OF MILLIONS of dollars are spent researching the problem of the origin of life at scores of eminent labs around the world. Every year, new results generate a great deal of excitement that scientists may finally be on the brink of solving the central mystery of biology. A never-ceasing drumbeat of stories appears in the press about every new idea. Often these are given outsized significance. Even the notion of panspermia is routinely resurrected as an exciting and somehow new idea. We want to believe that science has a firm grip on the central problem of biology. The reality, though, is that it is very difficult to say how close we are to understanding the answer to this most vexing of questions.

Yet the search for answers has already taught us an enormous amount about the world and the way it works. Since the Reinaissance, scientists involved in that search have transformed our understanding of biology. They have driven our first steps into the cosmos and spurred our exploration of the microscopic workings of molecular biology. Along the way, our vision of the universe and our place within it have been fundamentally reshaped.

The long saga to understand the origin of life may hold some subtler lessons as well. It may tell us something deeper about the nature of science, even something about our very nature as human beings. Most of the characters in the great saga to unravel the origin of life did not merely set out to answer a question. Many of them used science to prove or disprove a worldview.

Redi didn't drink snake venom just to make a point about the deadliness of snake venom. He was trying to show that reason was superior to superstition. The atheist supporters of John Needham did not look at his work and say, "This is good science." They looked into his microbial broth and saw a pathway around the need for God. The religious looked at Needham's broth in much the same way, and saw a threat to the existential meaning upon which they had centered their lives. Sidney Fox looked at his proteinoid microspheres and saw a vindication of his life's work, something he could never let go of, even when faced by a mountain of opposition. Nearly every scientist who attempts to explain purported microbial fossils—in ancient rocks or in meteorites—sees something different. These tiny structures often become the scientific equivalent of a Rorschach test.

Champions of experimentally derived knowledge can look at the same evidence and see something completely different, even diametrically opposed, because science and the scientist do not exist in a vacuum. They exist in a real world of constantly changing ideas and beliefs. Much has changed since Pasteur's time. Society is different. Religion is different. What we know or believe we know about the world is different. How humanity sees its relationship to that world is different. And consequently, so is what humanity sees in science.

This is evident in no field of science more than the study of life's origin. Most people cannot seem to simply disregard the question, cannot simply say, "I do not know," as people might once have responded to a question about why lightning occurs, or might now answer a question about the nature of dark matter. We may not know or claim to know the exact details of how life began, but we hold in our heads answers based on philosophy, religion, conjecture, even wishful thinking. We hold these answers because we have to, because the origin of life strikes at the very meaning of what it is to be alive. Scientists are no different. They often cling to their particular answer even in the face of contradictory evidence, even when, in cases like those of Sidney Fox and Charles Bastian, their intransigence means professional loss.

Yet science remains the best method for understanding the world. There is a process of scrutiny, a process of *provando e riprovando* in the spirit of the Medicis' Accademia del Cimento. In the end, the truth is often arrived at. Or, at least, something closer to the truth. Today, almost everyone would agree that flies come from eggs, not rotting meat. Pasteur was right about the existence of airborne germs. Bastian was wrong. One man's views became accepted as common sense; the other's were forgotten. We can say these things are true because, in the long run, science does work in the impartial fashion described by both Pasteur and Dawkins. Its history is filled with figures who chose evidence over their own beliefs or what may have been convenient. Darwin took no pleasure in his role as what he called the "Devil's Chaplain," pitted against pious men whom he respected and the church that he attended. He shared his science with the world at large only with great reluctance. Yet ultimately, he did. Our understanding of the natural world has grown exponentially since then.

In the end, science will weed out old falsehoods and will reveal new truths. But the path to such understanding might not always be as clear and straightforward as it appears, and the winners and losers might not always be immediately evident. Writing in *Scientific American* in 1954, biologist George Wald made that observation about Pasteur's great "triumph" in the spontaneous-generation debates of the late nineteenth century:

> It is no easy matter to deal with so deeply ingrained and common-sense a belief as that in spontaneous generation. One can ask for nothing better in such a pass than a noisy and stubborn opponent, and this Pasteur had in the naturalist Felix Pouchet, whose arguments before the French Academy of Sciences drove Pasteur to more and more rigorous experiments. When he had finished, nothing remained of the belief in spontaneous generation. We tell this story to beginning students of biology as though it represents a triumph of reason over mysticism. In fact it is very nearly the opposite. The reasonable view was to believe in spontaneous generation; the only alternative, to believe in a single, primary act of supernatural creation. There is no third position.

The history of science is filled with "losers" who clung to a conclusion despite the rejection of their peers. They all possessed a stubbornness that, for some, led to professional disgrace. Nevertheless, they took that road. On the surface, their inability to simply abandon their positions in the face of intense criticism or even contrary evidence smacks of hubris.

Their doggedness may serve a purpose. The naturalist Alexander von Humboldt once remarked that there are three phases of scientific discovery. The first is denial. The second is denial of importance. The third is crediting the wrong person. It takes a certain kind of fortitude to overcome the first step. Truly novel thinkers are often treated as crackpots. When proved wrong, history decrees that they remain crackpots. When proved right, history recasts them as visionary geniuses. The crackpots of the past may become the visionaries of the future.

THE SCIENTIFIC SEARCH for the origin of life continues. Meanwhile, we see that science, like history, tends to repeat itself. Every generation finds a new messenger with a definitive answer to the question, and every generation finds a new debate. Some bold scientist will push his or her Sisyphean answer up a hill only to find it rolling down again. Something will be seen in the lens of a microscope or in some test tube or fossil or rock, only to be reexamined, rescrutinized, and eventually reinterpreted. And in their search for answers, scientists often find themselves returning to the once-discarded solutions of the past. The debates of Needham and Spallanzani bear a striking resemblance to those between van Helmont and Redi, and between Pasteur and Bastian. During each epoch a definitive "victor" is proclaimed, only to see victory overturned by future discoveries.

For now, the "losers" have all but disappeared from textbook science. But science can have a short memory. Their sagas will be told again and differently, and we do not yet know how the tale will end. The forgotten may be resurrected. New scientists will take up where the old left off. We may still find answers in a discarded idea or discovery that is presently thought to have little, if any, relevance.

WHEN OR IF an explanation to the problem of life's origin is found, a solution capable of withstanding all the rigors of scientific scrutiny, we might find that the real answer we have been looking for continues to remain elusive. That is because there has always been a bigger question looming over the search for life's origins. It is the reason the debate has engendered such strong emotions and visceral reactions, and why the question has led so many scientists to throw scientific caution to the wind. For as human beings have searched for the *origin* of life, what they often seem to have been searching for is the *meaning* of life. That, perhaps, is something that science alone will never be able to answer.

APPENDIX: RECIPES FOR LIFE

Johannes van Helmont's Recipe for Mice

Place a dirty shirt or some rags in an open pot or barrel containing a few grains of wheat or some wheat bran, and in 21 days, mice will appear. Adult males and females will be present, and they will be capable of mating and reproducing more mice.

Henry Bastian's Recipes for Microbes

1. Boil a flask containing beef juice for 15 minutes, and then place it under vacuum and hermetically seal it. After 12 days the liquid will contain actively moving bacteria and several monads.

2. Boil a flask containing a rather weak infusion of beef, carrot, and turnip for 15 minutes, and then place it under vacuum and hermetically seal it. After 14 days the liquid will contain yeast-like cells.

3. Boil a flask containing a neutral solution of white sugar, ammonium tartrate, ammonium carbonate, and ammonium phosphate for 20 minutes, and then place it under vacuum and hermetically seal it. After 9 days the liquid will contain yeasts, bacteria, and monads.

4. Boil a flask containing a solution of ammonium oxalate and sodium phosphate for 20 minutes, and then place it under vacuum and hermetically seal it. After 61 days the liquid will contain fungal spores, as well as monads showing "tolerably active movements."

Sidney Fox's Recipe for Proteinoid Microspheres

Heat 10 grams of L-glutamic acid at 175°C–180°C until molten (about 30 minutes). Then add 10 grams of DL-aspartic acid and 5 grams of a mixture of the sixteen basic and neutral amino acids. Maintain the solution at 170 ± 2°C under an atmosphere of nitrogen for a few hours. During that time, considerable gas will have evolved, and the color of the liquid will have changed to amber. Rub the mixture vigorously with 75 milliliters of water, converting it to a yellow-brown granular precipitate. This material will form protocells, which should be capable of division and self-replication. Most of these materials can be obtained at your local health food store. We estimate the cost of this experiment to be about a hundred dollars.

Craig Venter's Recipe for a Cell

You will need a DNA synthesizer and a fairly sophisticated molecular biology laboratory. You will need to make, base pair by base pair, a complete bacterial genome, involving the construction of a chain of several million chemical linkages one by one. You will then need to insert this molecule, which, at today's prices, will cost you upwards of a million dollars, into a living bacterium. If you have placed the appropriate markers into the genome you made, you will be able to purify the synthetic descendants of your creation from the natural ones.

NOTES

Chapter 1

Carlo Rovelli's book *The First Scientist: Anaximander and His Legacy*, contains a wealth of information about this most underappreciated of Greek philosophers. Anaximander's views on the origin of life are discussed in Henry Osborn's *From the Greeks to Darwin*. Biographical details of Augustine are taken from Peter Brown's *Augustine of Hippo*.

5 **The Greeks have always received a lot of credit**: Patricia Fara provides a superbly comprehensive, non-Eurocentric history of science in *Science: A Four Thousand Year History*.

7 **He set his sights on the sun, the stars**: Anaximander's ideas on physics are discussed in David Park's *The Grand Contraption*.

8 **In Anaximander's scheme**: Osborn, *From the Greeks to Darwin*, 33–35.

10 **Just as the Galápagos would provide**: A beautifully written account of Aristotle's stay on Lesbos can be found in *Darwin's Ghosts*, by Rebecca Stott.

11 **"was devoted at all times to magic"**: Charles, *Chronicle of John*, 100.

12 **"who persist in applying their studies to a vain purpose"**: Lindberg, "Fate of Science," 22.

13 **Augustine also turned his inquisitive eye**: Augustine, *City of God*, 102.

13 **The policy was a result**: Fry, *Emergence of Life on Earth*, 20.

14 **Some forty years after *Antony and Cleopatra***: Cobb, *Generation*, 10.

Chapter 2

Francesco Redi's account of his experience with the Franciscans and their supposed wards against poison can be found in a letter to Athanasius Kircher that was included in Redi's 1687 book *Esperienze intorno*.

15 **The grand duke had a reputation:** The relationship between Ferdinando II and Redi is well elaborated in Harold Acton's *Last Medici*.

17 **"Doubt often wants to grow":** Redi, *Esperienze intorno*, 7.

19 **A native of Brussels:** *A Short History of Chemistry*, by J. R. Partington, contains a great summation of van Helmont's place in the history of chemistry.

20 **Many years later, he would write:** Ibid., 44.

20 **It involved mixing a sweaty shirt:** Cobb, *Generation*, 10.

21 **But Redi had an epiphany:** Redi's *Experiments on the Generation of Insects* contains a firsthand account of the events that led him to conduct his experiments refuting spontaneous generation.

23 **The official cause was "apoplexy":** Acton, *Last Medici*, 106–8.

23 **The new grand duke was his mother's child:** Cosimo III's religious extremism is covered by Acton in *The Last Medici* and by Christopher Hibbert in *The House of Medici*.

23 **A biographer would later describe Cosimo:** Acton, *Last Medici*, 112.

Chapter 3

A large selection of Antonie van Leeuwenhoek's letters is available under the title *The Collected Letters of Antoni van Leeuwenhoek*, although this collection still contains only a fraction of his huge body of correspondence. Clifford Dobell's *Antony van Leeuwenhoek and His "Little Animals"* offers a wealth of additional biographical information.

28 **"You are either a Spinozist":** Duquette, *Hegel's History of Philosophy*, 144.

30 **Latin and Greek were virtually mandatory:** Jonson, *Works of Ben Jonson*, 3:287.

31 **Simple lenses had been around:** The Roman emperor Nero was said to have watched gladiatorial combat using an emerald as a corrective lens. Between the eleventh and thirteenth centuries, "reading stones" were crafted in Italy to assist older people with declining eyesight. Tommaso da Modena's 1352 portrait of a bespectacled Cardinal Hugh de Provence is the first image of someone using a lens as a reading aid. Monks often used lenses to assist in illuminating manuscripts, and a Florentine manuscript from 1289 describes glass curved in such a way that it had "great advantage to old people with weak vision." On account of its oval shape, the writer called such glasses *lenti*, the Italian for lentils. from the Italian *lenticchia* ("lentil"). From *lenti* sprang the world "lens."

33 **By the time *Micrographia* appeared:** Van Leeuwenhoek's house became the subject of a Jan Vermeer painting, *A House in Delft*. It is notable for being the only painting that Vermeer ever set outside the confines of his own study.

35 Oldenburg was released after the threat: Dobell, *Antony van Leeuwenhoek and "His Little Animals"*, 39.

36 In 1673, the journal included a letter: Ibid., 41.

36 Huygens wrote that van Leeuwenhoek: Ibid., 43.

37 "I have no style, or, pen": Ibid.

37 Surely, Oldenburg wrote: Ibid., 42.

38 "The vermin only tease and pinch": D. F. Harris, "Antony van Leeuwenhoek the First Bacteriologist."

39 While examining his own saliva: Dobell, *Antony van Leeuwenhoek and "His Little Animals"*, 239.

39 In an old man who "never washed": Ibid.

39 In a letter to the Royal Society: Ibid., 243.

40 "It was just as impossible": H. Harris, *Things Come to Life*, 30.

41 He even convinced himself: Most of the smallest organisms that van Leeuwenhoek observed reproduced asexually by dividing in two in a process now known as binary fission.

42 In a 1692 essay on the state of microscopy: Bradbury, *Evolution of the Microscope*, 76.

42 One of his letters described a fit: Dobell, *Antony van Leeuwenhoek and "His Little Animals"*, 91.

43 On receipt, a clerk at the Royal Society: Ibid., 96.

43 The bequest was accompanied: Ibid., 97.

Chapter 4

Science historian Shirley Roe has written extensively on the pamphlet debate between Voltaire and John Needham, both in her chapter "Biology, Atheism, and Politics in Eighteenth-Century France" in the anthology *Biology and Ideology* and in her other academic works on the subject.

45 "It's better to go along with the stories": Park, *Grand Contraption*, 26.

47 In 1757, amid the reactionary climate: Parton, *Life of Voltaire*, 2:277.

47 "It is dangerous to be right": Moland, *Oeuvres complètes de Voltaire*, 14:73.

48 In a letter to his lifelong confidant: Becker and Becker, *Encyclopedia of Ethics*, 3:1771.

48 "Is it not the most absurd of all extravagances": Voltaire, *Works of Voltaire*, 273.

48 "Miracles," he said, "are very intelligible": Roe, "Voltaire versus Needham," 74.

49 "hold the Christian sect in horror": Gay, *Enlightenment*, 391.

52 "One can say that in a single apple pit": Malebranche, *De la recherche de la vérité*, 46–48.

53 "If one knew what all the parts": Roe, "Biology, Atheism, and Politics," 40.

53 He compared the process to that of "a clock": Pinto-Correia, *Ovary of Eve*, 1.

54 French author Bernard de Fontenelle: Broman, "Matter, Force and the Christian Worldview," 93.

54 "even the tiniest fibril": Jacob, *Logic of Life*, p. 76.

56 "I could hardly believe my eyes": Dawson, *Nature's Enigma*, 95.

56 When he presented a demonstration: Stott, *Darwin's Ghosts*, 96.

58 "My Phial swarmed with Life": H. Harris, *Things Come to Life*, 40.

59 He was convinced that these: Ibid., 42.

59 "Living and animation": Roe, "Biology, Atheism, and Politics," 45.

60 Buffon did something else in *Natural History*: The significance of Buffon's use of the word "reproduction" is discussed in detail in François Jacob's classic *The Logic of Life*.

60 "Needham has seen, has imagined": Roe, "Voltaire versus Needham," 77.

61 In letters to friends at home: Mitford, Nancy, *Voltaire in Love*, 23.

61 "The more I glimpse of this philosophy": Davidson, *Voltaire: A Life*, Kindle location 2058.

63 In a letter to Frederick: Hamel, *Eighteenth Century Marquise*, 370.

64 Maupertuis described Needham's experiment: Roe, "Voltaire versus Needham," 72.

64 "100,000 madmen of our species wearing hats": Voltaire, *Works of Voltaire*, vol. 33, 1829.

64 He wrote another satire, *Séance memorable*: Ibid.

65 "If God did not exist": Voltaire, *Oeuvres complètes de Voltaire*, 10:402.

66 "If I examine on the one hand": Roger, *Life Sciences in Eighteenth-Century French Thought*, 518.

66 "A watch," Voltaire said, "proves a watchmaker": Israel, *Enlightenment Contested*, 364.

66 "You had made small reputation": Roe, "Biology, Atheism, and Politics," 49.

68 "Men will always deceive themselves": D'Holbach, *System of Nature*, 11.

69 In a footnote, he invited readers: Ibid., 21.

69 Voltaire called it "a great moral sickness": Stott, *Darwin's Ghosts*, 157.

69 To a friend he wrote: Roe, "Voltaire versus Needham," 81.

69 "The world recoils in horror": Ibid., 83.

71 "I die adoring God": Espinasse, *Life of Voltaire*, 191.

71 He wrote a parody: Roe, "Voltaire versus Needham," 83.

Chapter 5

Andrew Crosse's second wife, Cornelia Crosse, published an account of his life under the title *Memorials, Scientific and Literary, of Andrew Crosse, the Electrician*. This work includes many of Crosse's own memoirs, as well as accounts from friends. James Secord's *Victorian Sensation* is a superb telling of the story behind the authorship of and associated controversy that greeted *Vestiges of the Natural History of Creation*. It should be required reading for anyone interested in the history of Victorian-era science.

73 **No right-minded architect:** Crosse and Crosse, *Memorials*, 153.
78 **"Hence without parent by spontaneous birth":** Nichols, *Romantic Natural Histories*, 129.
79 **The analogy people drew:** Andrew Crosse continues to draw interest because of the rather specious theory that Mary Wollstonecraft Shelley based her character Dr. Frankenstein on Crosse. The theory has even been the subject of a relatively recent book. Although Shelley did, in fact, attend a lecture Crosse had given on electricity, nothing prior to his 1836 experiment tied him to spontaneous generation, the creation of life, or anything else to do with the biological sciences. His famous experiment was stumbled upon purely by accident, and it happened several years after Shelley's book was written. Rather than Crosse influencing Shelley, it was she who influenced him, at least in the way he was later perceived.
79 **"a spark of being into a lifeless thing":** Shelley, *Frankenstein*, 34.
79 **"So easy it is to deceive oneself":** Whittaker, *History of the Theories of Aether and Electricity*, 69.
80 **A year later, Aldini electrically animated:** Lane, *Life Ascending*, 149.
81 **The first person known to have written:** William Gilbert, the personal physician of Queen Elizabeth I, was the first to systematically describe electrical phenomena. Gilbert was of the category of late-Renaissance scientists who had begun to break with the Greek classicists. He made a great show of his contempt for scientists who based everything on Aristotle's writings and would "toss off a few Latin words in the hearing of the ignorant rabble in token of their learning." Yet when he gave his phenomenon a name in 1600, he used the Latin word *electricus*, meaning "from amber." A century and a half later, *electricus* began to be replaced by the word "electricity."

Gilbert's reference appeared in the pages of his seminal work, *De magnete, magneticisque corporibus, et de magno magnete tellure*, published in

1600. His interest in magnetism was an outgrowth of his interest in astronomy. Gilbert was an adherent of Copernican theories of planetary motion, in which planets orbited the sun and not vice versa, and he came up with a rather brilliant theory for why this was so. He imagined the Earth—and all the heavenly bodies—to be giant magnets with their orbital paths based on magnetic fields. It was a remarkably prescient guess at a time before Newton had formulated his theory of gravity. Gilbert's contemporary, the German astronomer Johannes Kepler, arrived at many of the same conclusions.

82 **Still others called it an "imponderable fluid":** Tresch, *Romantic Machine*, 46.

83 **So eager were people:** Ibid.

84 **Priestley also was the first person:** Benjamin Franklin drew his friend Joseph Priestley into the field of electricity by posing to him a problem. At the time, scientists often used a kind of parlor trick to demonstrate the nature of electricity. All it required was an electrically charged metal can and a piece of cork tied to a string. The cork would be held close to the can and drawn to it. Once the can and the cork touched, the cork would itself become electrically charged, and it would be deflected away. Franklin noticed that if he placed the cork inside the can, the cork didn't move at all, which perplexed him. Priestley figured out that this lack of movement was due to the equal attraction of the cork from all sides of the can. More important, Priestley was able to deduce that electrical force followed the same mathematical behavior Newton had used to describe gravity: diminishing as the inverse square of the distance between two interacting bodies.

Priestley's book *The History and Present State of Electricity* marked the first appearance of the story of Franklin's kite flying during an electrical storm, which Priestley almost certainly heard from the famously self-promoting Franklin himself. The image of the sagacious statesman bravely flying a kite in a rainstorm became the iconic depiction of one of America's most colorful characters. Whether or not it actually happened remains an open question.

85 **Crosse was a loner at heart:** Crosse and Crosse, *Memorials*, 32.

85 **Many years later, he would say:** Ibid., 33.

86 **Using an electric current:** Humphry Davy's contributions are beautifully recounted in *The Age of Wonder*, by Richard Holmes.

88 **The novelist and future prime minister:** Secord, *Victorian Sensation*, 10.

89 **He sent Crosse a personal letter:** Secord, "Curious Case," 472.

90 **In print, he was branded:** Ibid.

Chapter 6

Darwin: The Life of a Tormented Evolutionist, by Adrian Desmond and James Moore, is the seminal biography of Charles Darwin. Darwin's views on the origin of life have been elaborated by the scientists Juli Peretó, Jeffrey L. Bada, and Antonio Lazcano, and the historian of science James Strick.

93 **FitzRoy described the volcanic beach:** Desmond and Moore, *Darwin*, 169.

94 **In the three weeks before he arrived:** Ibid.

95 **They encountered a group of Spanish whalers:** C. Darwin, *Voyage of the Beagle*, 381.

97 **If these birds indeed turned out to be:** C. Darwin, "Darwin's Ornithological Notes," 262.

98 **"We seem to be brought somewhat near":** C. Darwin, *Voyage of the Beagle*, 400.

100 **"One might really fancy":** Ibid., 402.

102 **His ideas earned him a place:** The horrible conditions of these workhouses would one day become infamous, but at the time, they were popular with the progressive reformers who dominated the Whig Party. The Poor Laws were passed with mostly Whig support and opposed both by Tory conservatives who favored traditional charity and by working-class radicals. Though Malthusian economics was cited by supporters of the laws, Malthus himself opposed them.

102 **Darwin would later write:** C. Darwin, *Autobiography*, 98–99.

102 **Joseph Priestley, himself no orthodox follower:** Priestley, "Observations and Experiments," 128.

105 **"The watch must have had a maker":** Paley, *Natural Theology*, 1.

105 **When Darwin read Paley's book:** C. Darwin, *Autobiography*, 51.

105 **"I fear there are but small hopes":** Desmond and Moore, *Darwin*, 191.

106 **She wrote to him afterward:** Brown, *Darwin's Origin of Species*, 46.

108 **"We went a little into society":** C. Darwin, *Charles Darwin*, 37.

110 **His son later noted the book's "simplicity":** Brown, *Darwin's Origin of Species*, 67–68.

110 **In its stead was nature:** C. Darwin, *Annotated Origin*, 84.

110 **"If you be right":** C. Darwin, *Correspondence*, 7:379.

110 **"There is grandeur in this view of life":** The changes to this important paragraph in *Origin* has been traced at Darwin Online (http://darwin-online.org .uk/Variorum/1859/1859-490-c-1860.html).

112 **It smacked of cowardice:** As we write this book, the Institute for Creation Research, a Christian group with a mission to undermine science that contradicts biblical scripture, maintains a website with a section on the origin of life that notes Darwin's failure to address the question in *Origin*, "even though it's the title of his book."

113 **"The doctrines of the *generatio spontanea*":** Rupke, *Richard Owen*, 173.

113 **"Is there a fact, or a shadow of a fact":** Peretó, Bada, and Lazcano, "Charles Darwin and the Origin of Life," 399.

113 **But when it came to his choice of words:** Ibid.

113 **"I have long regretted":** Ibid.

113 **It would be, he guessed:** Peretó, Bada, and Lazcano, "Charles Darwin and the Origin of Life," 401.

114 **"At the present day":** Ibid.

Chapter 7

Two excellent sources of information on Pasteur—although from radically different perspectives—are Patrice Debré's biography (*Louis Pasteur*) and Gerald L. Geison's critical reappraisal of the scientist's work and life (*The Private Science of Louis Pasteur*). Geison's otherwise excellent account falls a little short on Pasteur's conflict with Pouchet, failing to adequately point out that, ultimately, Pasteur was right about spontaneous generation, at least in the observable cases presented by Pouchet, Bastian, and those who came before them. Bastian's story is at the center of science historian James Strick's *Sparks of Life: Darwinism and the Victorian Debates over Spontaneous Generation*, which does a good job of showing how the conflict between miasmatic theory and germ theory became wrapped up in the contentious debate over spontaneous generation.

118 **Pasteur began his address:** Pasteur, "On Spontaneous Generation."

119 **"We simply take a drop of sea-water":** Geison, *Private Science of Louis Pasteur*, 111.

119 **"Mightn't matter, perhaps, organize itself?":** Ibid.

120 **In Pasteur's later years:** The importance Pasteur placed on spontaneous generation is corroborated in Debré's *Louis Pasteur*.

122 **When Lamarck died penniless and nearly blind:** Cuvier, G. "Biographical Memoir of M. de Lamarck," 1.

123 **To a friend, Goethe likened the debate:** Appel, *Cuvier-Geoffroy Debate*, 1.

124 **The reason for the phenomenon:** The origin of homochirality is still a topic of scientific debate. The best guess to date is that in order for large molecules like proteins or nucleic acids to function properly, homochirality is neces-

sary. The initial bias toward one or the other may have been inherited from the pre-solar-system outer-space environment, where circularly polarized light from neutron stars and supernovas selectively formed one or the other orientation of the molecule. Homochirality should not be confused with the distinction between organic and nonorganic molecules. Organic molecules can be homochiral, but they don't *need* to be.

125 **Soon, his wife was writing her father:** Debré, *Louis Pasteur*, 87.

126 **Pasteur wrote to a colleague:** Ibid., 150.

127 **Darwin, who knew nothing of Royer:** Prum, "Charles Darwin's First French Translations," 392.

128 **In their judgment, they wrote:** Geison, *Private Science of Louis Pasteur*, 125.

129 **One, the anatomist Thomas Huxley:** Jensen, "X Club," 64.

130 **That night, they decided to form:** Ibid.

131 **It was held at Oxford:** Desmond and Moore, *Darwin*, 493.

135 **In a letter to Wallace:** Peretó, Bada, and Lazcano, "Charles Darwin and the Origin of Life," 398.

137 **From the south of France:** Debré, *Louis Pasteur*, 179.

141 **"We are told, indeed":** Brieger, *Medical America*, 286.

143 **"Do you know why it is so important":** Debré, *Louis Pasteur*, 300.

143 **In a vicious review of *The Beginnings of Life*:** Strick, *Sparks of Life*, 101.

144 **"To our mind the position is quite unchanged":** "Origin of Life," *Lancet*, 970.

145 **"Though no evidence worth anything":** F. Darwin, *More Letters of Charles Darwin*, 171.

Chapter 8

JBS: The Life and Work of JBS Haldane, by Ronald Clark, is a marvelous biography of a colorful figure whose significance has largely been forgotten. Biographical material on Alexander Oparin is harder to come by. William Schopf includes a nice firsthand portrait of the Russian scientist in *Cradle of Life*. Loren Graham's works on Soviet science (*Science in Russia and the Soviet Union* and *Science, Philosophy, and Human Behavior in the Soviet Union*) continue to stand out as the most authoritative record of that sad period in Russian history.

147 **A Scottish geologist by the name of James Hutton:** Dean, *James Hutton and the History of Geology*, 262.

148 **"Though speculations concerning":** "British Association—Leicester Meeting," 135.

149 **"Those who lived in close contact with him":** Clark, *JBS*, 45.

150 After injuring his forehead: Ibid., 13.

151 They were exceedingly unpredictable: Ibid., 36.

151 He later recalled that he "thought it important": Ibid., 37.

151 Sir Douglas Haig, commander: Ibid.

153 Haldane's admirer Arthur C. Clarke: Clarke, Foreword to *What I Require from Life*, ix.

153 Either an act of abiogenesis: Haldane, "The Origin of Life," 6.

154 On the seventy-fifth anniversary: Hyman and Brangwynne, "In Retrospect," 524.

156 Later, when he became a professor: Schopf, *Cradle of Life*, 112.

156 Bakh found himself gradually elevated: Communist dominance of Russian scientific institutions was not as quick as one might imagine. It would be a number of years until those institutions came under the kind of totalitarian control that characterized the Stalinist period. Not until 1929 did the Russian Academy of Sciences admit its first communists, a group that included Bakh and Oparin.

158 Kelvin postulated that the Earth: A deeper elaboration of Kelvin's methods can be found in "Kelvin, Perry and the Age of the Earth," by Philip C. England, Peter Molnar, and Frank Richter.

158 By 1897, he had settled on: Kelvin, *Mathematical and Physical Papers*, 5:215.

161 "If the theory [evolution] be true": Schopf, *Cradle of Life*, 15.

163 In 1891, Walcott wrote: Walcott, "Pre-Cambrian Rocks," 594.

167 The son of an illiterate peasant farmer: The *Pravda* article and much of the biographical information can be found in Zhores Medvedev's *The Rise and Fall of T. D. Lysenko*.

167 "If one is to judge a man": Medvedev, *Rise and Fall of T. D. Lysenko*, 11.

169 "If you had been there during those years": Graham, *Science in Russia and the Soviet Union*, p. 276.

169 "Sixty years in socks is enough": DeJong-Lambert, *Cold War Politics of Genetic Research*, 150.

170 He went on to say: Clark, *JBS*, 294.

170 "I suppose that Oparin and I": Ibid., 286.

Chapter 9

Detailed biographical treatments of many of the characters in this chapter, including Harold Urey, Stanley Miller, and Sidney Fox, can be found in a series of books entitled *Men of Space*. The series was authored in the 1960s by a former Hollywood actress and personality named Shirley Thomas, billed on the book jackets as "The First Lady of Space." James Strick's article "Creating a Cosmic Discipline," from the

Journal of the History of Biology, is an excellent account of the events leading to the establishment of NASA's exobiology program, as well as the program's early years.

173 **"If God did not do it this way"**: Bada and Lazcano, "Stanley Miller's 70th Birthday," 109.

174 **In 1952, Teller left Chicago**: In 1946, Teller became the first scientist to conceive of an atomic bomb that used hydrogen fusion to magnify its destructive power. After the Soviet Union's successful test of an atomic bomb in 1949, President Truman had started to steer money into research for Teller's weapon—what would come to be called the hydrogen bomb. The project initially met a lot of resistance from other nuclear scientists, who questioned Teller's assumptions. By 1952, however, advances in the understanding of nuclear physics had changed many minds in the scientific community. There were also concerns in the intelligence community that the Soviets were working on their own version of such a bomb, and the Defense Department wanted to jump-start its program. Teller took charge of the secret weapon project at the University of California's Radiation Laboratory at Lawrence Livermore. Soon he was named director of the US primary nuclear weapons development facility at Los Alamos. By the end of 1952, the United States had tested its first nuclear weapon based on Teller's design. A year later, the Soviets tested their own hydrogen bomb.

175 **Calvin was one of the world's greatest authorities**: Of all the metabolic functions of living organisms, photosynthesis is one of the most complex. Looking backward from a modern evolutionary vantage point, it's easy to see plants, bereft of nervous systems, as less "advanced" than animals. This was the view held by early evolutionists like Huxley, who thought the first organisms were likely some form of plantlike life akin to algae. Coming from a background in plant biology, Alexander Oparin saw it the opposite way. Fermenting microbes were the more basic organisms and must have appeared first. Plants, with their sophisticated ability to absorb energy from the sun, were more complex and must have evolved later. Oparin's view has come to be generally accepted in the scientific community.

179 **An editorial in the *New York Times***: "Life and a Glass Earth," *New York Times*.

179 ***Time* magazine reported**: "Science: Semi-Creation," *Time*.

180 **Urey told Miller he had to make up his own mind**: Lazcano and Bada, "Stanley L. Miller (1930–2007)," 379.

182 **After the results of the Miller-Urey experiment**: Schopf, *Cradle of Life*, 127.

182 **Later in the winter of 1957:** Lederberg recounted his meeting with Haldane in Calcutta in great detail in an article he wrote for the *Journal of Genetics* entitled "Sputnik + 30."

184 **As Carl Sagan later wrote:** Strick, "Creating a Cosmic Discipline," 135.

Chapter 10

A Man on the Moon, by Andrew Chaikin, stands out as the finest account of the *Apollo* missions. Mathew Ridley's *Francis Crick: Discoverer of the Genetic Code* is an excellent account of how the genetic machinery of living cells was elaborated. Ridley's popular book *Genome* also delves into the discoveries that paved the way for what would become a monumental advance in our understanding of biology.

188 **"The surface appears to be":** Chaikin, *Man on the Moon*, 208.

189 **Aldrin called it "magnificent desolation":** Ibid., 211.

190 **An elaborate quarantine center:** Gary McCollum and Donald Bogard's interview for NASA's oral history project goes into the precautions at length.

192 **"You are going to make a choice":** "Fox," *Mobile Register*.

193 **As Fox would later write:** S. W. Fox, "Apollo Program and Amino Acids," 46.

193 **He was looking for clues:** Ibid.

197 **"Fox, all the problems of life":** Strick, "Creating a Cosmic Discipline," 154.

197 **He returned to California:** "Sidney W. Fox," *Los Angeles Times*.

198 **His experiment, he claimed:** S. W. Fox, Harada, and Kendrick, "Production of Spherules."

200 **"I have never seen Francis Crick":** Watson, *Double Helix*, 7.

200 **And although his grandfather:** Ridley, "Crick and Darwin's Shared Publication," 244.

200 **He was rescued from this fate:** R. Alexander and Stevens, "Obituary: Francis Crick."

202 **Joshua Lederberg would one day:** Lederberg, "Transformation of Genetics by DNA."

202 **"We have as yet no actual knowledge":** Muller, "Development of the Gene Theory," 95–96.

203 **A blunt American with a crew cut:** Watson, *Double Helix*.

203 **Later, in *The Double Helix*:** Ibid., 14.

203 **"the product of an unsatisfied mother":** Watson, *The Double Helix*, p. 17.

204 **As early as 1927, the Soviet scientist Nikolai Koltsov:** "Consequences of Political Dictatorship," *Nature Reviews. Genetics*.

204 **The structure, Watson and Crick wrote:** Watson and Crick, "Molecular Structure of Nucleic Acids."

206 **Though it carried genetic information:** Ridley, *Francis Crick*, 171.

Chapter 11

Kathy Sawyer's *The Rock from Mars* is a thorough telling of the story behind ALH84001, although the excitement that still surrounded the meteorite at the time her book was written has largely faded. Peter Ward's *Life as We Do Not Know It* is a good scientific take on the question of life in space.

211 **She noticed that the rock:** Score's initial observations of ALH84001 can be read at http://curator.jsc.nasa.gov/antmet/samples/petdes.cfm?sample= ALH84001.

212 **But the technology used:** Today, the number of meteorites identified as Martian has grown to 134, including four actually observed dropping from the sky. One that fell in Egypt in 1911 was reported to have struck and killed a farmer's unfortunate dog. Since the planets are relatively close neighbors in the solar system, it would not be surprising for Martian rocks to be common on Earth. Some scientists have estimated that half a ton of Martian rocks fall onto the Earth each year. Of all the Martian meteorites discovered to date, half have been found in the ice fields of Antarctica.

212 **In any event, over the next several years:** Fry, *Emergence of Life on Earth*, 221.

212 **"But why, some say, the moon?":** Video and text of President Kennedy's speech at Rice University is available at http://er.jsc.nasa.gov/seh/ricetalk.htm.

216 **"Today Rock 84001 speaks to us":** The full text of Clinton's statement is available at http://www2.jpl.nasa.gov/snc/clinton.html.

216 **"We must regard it as probable":** Kelvin, "Presidential Address," 202.

217 **The clamor of interest:** In 1965, a different sample of Orgueil was found to have been tampered with. In 1864, someone had inserted plant fragments into the rock and then used coal and glue to try to make it appear that the plant was embedded in the meteorite. Rather than a hoax to create the impression of extraterrestrial life, historians believe the tampering was an attempt to undercut Pasteur's work on disproving spontaneous generation.

217 **In 1962, they reported their findings:** Nagy, Claus, and Hennessy, "Organic Particles Embedded in Minerals."

218 **Urey, ever the open-minded skeptic:** The controversy about Orgueil has never really gone away. Since 2000, former NASA engineer Richard Hoover has published a number of papers claiming microscope-image evidence of

life in the form of tiny bacterium-like structures in the Orgueil and other carbonaceous meteorites. The last of these articles was published in 2011 in the *Journal of Cosmology*. The claims have been met with a variety of criticisms by the scientific community and produced some blowback for NASA. The agency had not endorsed Hoover's work, though it was widely reported to have done so. James Randi, the popular debunker of all things paranormal and pseudo-scientific, subsequently awarded Hoover and the *Journal of Cosmology* his not-so-coveted "Pigasus Award" in the category of "Scientist" (Plait, "2011 JREF Pigasus Awards").

218 **"The study of carbonaceous meteorites":** Nagy and Lynch, "Life-like Forms in Meteorites," 606.

219 **He had his doubts about specific claims:** Gould, Stephen, "Life on Mars? So What?" *New York Times*, August 11, 1996.

221 **After two years of review:** Kerr, R. "Requiem for Life on Mars?"

Chapter 12

The New Foundations of Evolution, by Jan Sapp, is a detailed history of the modern study of microbial evolution, and contains a thorough account of Carl Woese's discovery of the archaea.

229 **"By deducing rather ancient ancestor sequences":** Letter from Carl Woese to Francis Crick, June 24, 1969, copy provided by George Fox.

232 **"The archaea are related to us":** Bult et al., "Complete Genome Sequence."

233 **Reflecting on his dramatic vindication:** Morell, "Microbiology's Scarred Revolutionary." Ernst Mayr remained one of the few prominent holdouts to the new tree of life, even though he and Woese had come to change places in the opinions of most scientists. Unable to bring himself to accept the finding he had so stringently opposed, Mayr died in 2005 still insisting that Woese was wrong.

235 **By swapping information freely:** We now know that viruses may also play an important role in fostering evolutionary change. Traditionally, viruses have been seen as nothing more than opportunists looking for a free lunch in the form of a host in which replicate. But when viruses move to a new host, they may bring pieces of DNA from their last host with them. Provided the host survives the infection, this DNA may become part of the host's new genetic complement and be passed on by binary fission to its progeny. Thus, viruses may play an important role in accelerating the spread of genetic diversity.

236 **Woese called horizontal gene transfer:** Woese, "Evolving Biological Organization," 106.

237 **An organic chemist before:** Wächtershäuser's brother-in law, interestingly, is George Fox, Woese's collaborator on the discovery of archaea. Fox introduced Wächtershäuser to Woese.

Chapter 13

Biographical details on Jack Szostak are drawn mostly from the autobiographical sketch he wrote for the Nobel Committee and from an interview conducted at his lab in Boston.

242 **"The first stage of evolution proceeds":** Gilbert, "RNA World."

245 **And though, like Henry Bastian and Francis Crick:** Szostak told us in an interview that if he were a young student today, with the advances in technology enabling closer study of the way the brain actually works, he would devote himself to studying the phenomenon of consciousness. He predicted that science in the twenty-first century would be dominated by the unraveling of this question.

246 **Venter and his team encoded:** Angier, "Peering over the Fortress."

248 **This made perfect sense:** Spiegelman's results raise the interesting question of why not all genomes eventually shrink down to the smallest size possible. The conditions in Spiegelman's experiments were very favorable for the survival of shorter and shorter sequences. In the real world, viral genome have to keep the information for making the polymerase and the other various proteins that enable it to infect its host; thus there is a constant tension between being able to make copies faster and at the same time keeping all of the information that is necessary to propagate both. There is therefore a size limit below which the virus would become nonfunctional and noninfectious, and thus go extinct.

Epilogue

253 **In his famous lecture at the Sorbonne:** Pasteur, "On Spontaneous Generation."

253 **Roughly a century and a half later:** Dawkins's monologue prior to his interview of Craig Venter can be seen on YouTube, at https://www.youtube.com/watch?v=3E25jgPgmzk.

254 **"Score one for Francis":** Ridley, *Francis Crick*, 208.

254 **In 1996, Pope John Paul II alluded to:** John Paul II, "Truth Cannot Contradict Truth."

257 **"It is no easy matter to deal with":** Wald, "Origin of Life," 45–46.

Appendix

259 **Johannes van Helmont's Recipe for Mice:** Cobb, *Generation*, 10.

259 **Henry Bastian's Recipes for Microbes:** Bastian, *Evolution and the Origin of Life*.

260 **Sidney Fox's Recipe for Proteinoid Microspheres:** S. W. Fox and Harada, "Thermal Copolymerization."

260 **Craig Venter's Recipe for a Cell:** Gibson et al., "Creation of a Bacterial Cell."

BIBLIOGRAPHY

Acton, H. *The Last Medici*. London: Thames & Hudson, 1932.

Alexander, D. R., and R. L. Numbers, eds. *Biology and Ideology from Descartes to Dawkins*. Chicago: University of Chicago Press, 2010.

Alexander, R., and C. F. Stevens. "Obituary: Francis Crick (1916–2004)." *Nature*, 430 (2004): 845–47.

Anderson, D. "Still Going Strong: Leeuwenhoek at Eighty." *Antonie van Leeuwenhoek* 106 (2014): 3–26.

Angier, N. "Peering over the Fortress That Is the Mighty Cell." *New York Times*, May 31, 2010.

Appel, T. *The Cuvier-Geoffroy Debate*. Oxford: Oxford University Press, 1987.

Augustine. *City of God*, vol. 2. New York: Doubleday, 1958.

Bada, J. L., and A. Lazcano. *Stanley L. Miller 1930–2007: A Biographical Memoir*. National Academy of Sciences, 2012.

Bada, J. L., and A. Lazcano. "Stanley Miller's 70th Birthday." *Origins of Life and Evolution of the Biosphere* 30 (2000): 107–12.

Balme, D. M. "Development of Biology in Aristotle and Theophrastus: Theory of Spontaneous Generation." *Phronesis* 7, no. 1 (2009): 91–104.

Bastian, H. C. *Evolution and the Origin of Life*. London: Macmillan, 1874.

Bastian, H. C. "Reply to Professor Huxley's Inaugural Address at Liverpool on the Question of the Origins of Life." *Nature* 2 (1870): 410–13.

Becker, L. C., and C. B. Becker. *Encyclopedia of Ethics*, vol. 3. New York: Taylor and Francis, 1992.

Bradbury, S. *The Evolution of the Microscope*. Oxford: Pergamon, 1967.

Brieger, G. *Medical America in the Nineteenth Century*. Baltimore: Johns Hopkins University Press, 1972.

"The British Association—Leicester Meeting." *Current Science* 2 (1933): 135–44.

Broman, T. H. "Matter, Force and the Christian Worldview in the Enlightenment." In *When Science and Christianity Meet*, edited by D. C. Lindberg and R. L. Numbers, 85–110. Chicago: University of Chicago Press, 2003.

Brown, P. *Augustine of Hippo*. London: Faber and Faber, 1967.

Browne, J. *Darwin's Origin of Species*. New York: Grove Press, 2006.

Bryan, W. J. *William Jennings Bryan's Last Message*. New York: Fleming H. Revel, 1925.

Bult, C. J., O. White, G. J. Olsen, L. Zhou, R. D. Fleischmann, G. G. Sutton, J. A. Blake, et al. "Complete Genome Sequence of the Methanogenic Archaeon, *Methanococcus jannaschii*." *Science* 273, no. 5278 (1996): 1058–73.

Burkhardt, F. H. "Darwin and the Copley Medal." *Proceedings of the American Philosophical Society* 145, no. 4 (2001): 510–15.

Campbell, N., and S. Miller. "A Conversation with Stanley Miller." *American Biology Teacher* 51, no. 6 (1989): 349–53.

Chaikin, A. *A Man on the Moon: The Voyages of the Apollo Astronauts*. New York: Penguin, 1994.

Chambers, R. *Vestiges of the Natural History of Creation and Other Evolutionary Writings*. Chicago: University of Chicago Press, 1994.

Charles, R. H., trans. *The Chronicle of John, Bishop of Nikiu*. Oxford: Oxford University Press, 1916.

Clark, R. *JBS: The Life and Work of JBS Haldane*. New York: Coward-McCann, 1968.

Clarke, A. Foreword to *What I Require from Life*, by J. B. S. Haldane, ix–xii. Oxford: Oxford University Press, 2009.

Cobb, M. *Generation: The Seventeenth-Century Scientists Who Unraveled the Secrets of Sex, Life and Growth*. New York: Bloomsbury, 2006.

The Collected Letters of Antoni van Leeuwenhoek. Lisse, Netherlands: Swets & Zeitlinger, 1996.

"The Consequences of Political Dictatorship for Russian Science." *Nature Reviews. Genetics* 2 (2001): 723–29.

Colp, R., Jr. "'Confessing a Murder': Darwin's First Revelations about Transmutation." *Isis* 77, no. 1 (1986): 7.

Crick, F., and L. E. Orgel. "Directed Panspermia." *Icarus* 19 (1973): 341–46.

Crosse, A., and C. A. H. Crosse. *Memorials, Scientific and Literary, of Andrew Crosse, the Electrician*. London: Longman, Brown, Green, Longmans, & Roberts, 1857.

Cuvier, G. "Biographical Memoir of M. de Lamarck." *Edinburgh New Philosophical Journal* 20, No. 29 (1836): 1–22.

Darwin, C. *The Annotated Origin*, annotated by J. T. Costa. Cambridge, MA: Harvard University Press, 2009.

Darwin, C. *The Autobiography of Charles Darwin*. New York: W.W. Norton, 1958.

Darwin, C. *Charles Darwin: His Life Told in an Autobiographical Chapter, and in a Seclected Series of His Published Letters*, edited by F. Darwin. London: John Murray, 1908.

Darwin, C. *The Correspondence of Charles Darwin*. Vol. 7, *1858–1859*. Cambridge: Cambridge University Press, 1991.

Darwin, C. "Darwin's Ornithological Notes." *Bulletin of the British Museum (Natural History)* 2, no. 7 (1963): 201–78.

Darwin, C. *The Voyage of the Beagle*. New York: P. F. Collier & Son, 1909.

Darwin, F., ed. *More Letters of Charles Darwin*, vol. 2. New York: Appleton, 1903.

Davidson, I. *Voltaire: A Life*. London: Profile Books, 2010. Kindle edition.

Dawson, V. P. *Nature's Enigma: The Problem of the Polyp in the Letters of Bonnet, Trembley and Réaumur*. Philadelphia: American Philosophical Society, 1987.

Dean, D. R. *James Hutton and the History of Geology*. Ithaca, NY: Cornell University Press, 1992.

Debré, P. *Louis Pasteur*. Baltimore: Johns Hopkins University Press, 1994.

DeJong-Lambert, W. *The Cold War Politics of Genetic Research: An Introduction to the Lysenko Affair*. Dordrecht: Springer, 2012.

De Kruif, P. *Microbe Hunters*. New York: Harcourt, 1926.

Desmond, A., and J. Moore. *Darwin: The Life of a Tormented Evolutionist*. New York: W. W. Norton, 1991.

D'Holbach, Baron. *The System of Nature: or, the Laws of the Moral and Physical World*. New York: G. W. & A. J. Matsell, 1835.

Dimension, T. V. "The Victorian Conflict between Science and Relgion: A Professional Dimension." *Isis* 69, no. 3 (1978): 356–76.

Dobell, C. *Antony van Leeuwenhoek and His "Little Animals": A Collection of Writings by the Father of Protozoology and Bacteriology*. New York: Dover, 1960.

Dronamraju, K. R., ed. *Haldane and Modern Biology*. Baltimore: Johns Hopkins University Press, 1968.

Duquette, D. A. *Hegel's History of Philosophy: New Interpretations*. Albany: State University of New York Press, 2003.

England, P. C., P. Molnar, and F. M. Richter. "Kelvin, Perry and the Age of the Earth." *American Scientist* 95 (2007): 342–49.

Espinasse, F. *Life of Voltaire*. London: Walter Scott, 1892.

Falk, R. "Mendel's Impact." *Science in Context* 19 (2006): 215–36.

Fara, P. *Science: A Four Thousand Year History*. Oxford: Oxford University Press, 2009.

Farley, J. *The Spontaneous Generation Controversy from Descartes to Oparin*. Baltimore: Johns Hopkins University Press, 1977.

Figes, O. *A People's Tragedy: The Russian Revolution 1891–1924*. New York: Penguin, 1996.

"Fox." *Mobile Register*, August 13, 1998.

Fox, R. *The Savant and the State: Science and the Cultural Politics in Nineteenth Century France*. Baltimore: Johns Hopkins University Press, 2012.

Fox, S. W. "The Apollo Program and Amino Acids: On the Origin of Life." *Bulletin of the Atomic Scientists* 29, no. 10 (1973): 46–51.

Fox, S. W. *The Emergence of Life*. New York: Basic Books, 1988.

Fox, S. W., and K. Harada. "Thermal Copolymerization of Amino Acids to a Product Resembling Protein." *Science* 128 (1958): 1214.

Fox, S. W., K. Harada, and J. Kendrick. "Production of Spherules from Synthetic Proteinoid and Hot Water." *Science* 129 (1959): 1221–23.

Frangenberg, T. "A Private Homage to Galileo. Anton Domenico Gabbiani's Frescoes in the Pitti Palace." *Journal of the Warburg and Courtauld Institutes* 59 (1996): 245–73.

Friend, T. *The Third Domain: The Untold Story of the Future of Biochemistry*. Washington, DC: Joseph Henry Press, 2007.

Fry, Iris. *The Emergence of Life on Earth*. New Brunswick, NJ: Rutgers University Press, 2000.

Gay, P. *The Enlightenment: An Interpretation*. New York: Knopf, 1967.

Geison, G. L. *The Private Science of Louis Pasteur*. Princeton, NJ. Princeton University Press, 1995.

Gibson, D. G., J. I. Glass, C. Lartigue, V. N. Noskov, R.-Y. Chuang, M. A. Algire, G. A. Benders, et al. "Creation of a Bacterial Cell Controlled by a Chemically Synthesized Genome." *Science* 329, no. 5987 (2010): 52–56.

Gilbert, W. "The RNA World." *Nature* 319 (1986): 618.

Gould, S. "Life on Mars? So What?" *New York Times*, August 11, 1996.

Graham, L. *Science in Russia and the Soviet Union: A Short History*. Cambridge: Cambridge University Press, 1993.

Graham, L. *Science, Philosophy, and Human Behavior in the Soviet Union*. New York: Columbia University Press, 1987.

Haldane, J. B. S. *On Being the Right Size and Other Essays*. Oxford: Oxford University Press, 1985.

Haldane, J. B. S. "The Origin of Life." *Rationalist Annual* (1929): 3–10.

Hamel, F. *An Eighteenth Century Marquise: A Study of Émilie du Châtelet and Her Times.* London: Stanley Paul, 1910.

Harris, D. F. "Antony van Leeuwenhoek the First Bacteriologist." *Scientific Monthly* 12, no. 2 (1921): 150–60.

Harris, H. *Things Come to Life: Spontaneous Generation Revisited.* Oxford: Oxford University Press, 2002.

Hawgood, B. J. "Francesco Redi (1626–1697): Tuscan Philosopher, Physician and Poet." *Journal of Medical Biography* 11 (2003): 23–34.

Henry, F. "Rue Cuvier, rue Geoffroy-Saint-Hilaire, rue Lamarck: Politics and Science in the Streets of Paris." *Nineteenth Century French Studies* 35, no. 3 (2007): 513–25.

Henry, J. "Religion and the Scientific Revolution." In *The Cambridge Companion to Science and Religion*, edited by P. Harrison, 39–58. Cambridge: Cambridge University Press, 2010.

Hibbert, C. *The House of Medici: Its Rise and Fall.* New York: HarperCollins, 1974.

Hofstadter, D. *The Earth Moves: Galileo and the Roman Inquisition.* New York: W. W. Norton, 2009.

Holmes, R. *The Age of Wonder: How the Romantic Generation Discovered the Beauty and Terror of Science.* New York: Pantheon, 2008.

Hoover, R. B. "Fossils of Cyanobacteria in CI1 Carbonaceous Meteorites." *Journal of Cosmology* 13 (2011).

Hyman, T., and C. Brangwynne. "In Retrospect: The Origin of Life." *Nature* 491 (2012): 524.

Israel, J. *Enlightenment Contested.* Oxford: Oxford University Press, 2006.

Jacob, F. *The Logic of Life.* Princeton, NJ: Princeton University Press, 1973.

Jensen, J. V. "The X Club: Fraternity of Victorian Scientists." *British Journal for the History of Science* 5, no. 1 (1970): 63–72.

John Paul II. "Truth Cannot Contradict Truth." Message delivered to the Pontifical Academy of Sciences, October 22, 1996.

Jonson, Ben. *The Works of Ben Jonson*, vol. 3. London: Chatto & Windus, 1910.

Joravsky, D. *The Lysenko Affair.* Cambridge, MA: Harvard University Press, 1970.

Kamminga, H. *Studies in the History of Ideas on the Origin of Life from 1960.* London: Chelsea College, University of London, 1980.

Kaufman, M. *First Contact: Scientifc Breakthroughs in the Hunt for Life beyond Earth.* New York: Simon & Schuster, 2011.

Keenan, M. E. "St. Augustine and Biological Science." *Osiris* 7 (1939): 588–608.

Kelvin, William Thomson, Baron. *Mathematical and Physical Papers*, vol. 5. Cambridge: Cambridge University Press, 1911.

Kelvin, William Thomson, Baron. "Presidential Address to the British Association, Edinburgh, 1871." In *Popular Lectures and Addresses*, 132–205. London: Macmillan, 1894.

Kerr, R. "Requiem for Life on Mars? Support for Microbes Fades." *Science* 282, no. 5393 (1998): 1398.

Knoll, A. H. *Life on a Young Planet: The First Three Billion Years of Evolution on Earth*. Princeton, NJ: Princeton University Press, 2003.

Kohler, R. E., Jr. "The Enzyme Theory and the Origin of Biochemistry." *Isis* 64, no. 2 (1973): 181–96.

Kursanov, A. L. "Sketches to a Portrait of A. I. Oparin." Lecture presented at the International Symposium "Biochemistry of the 21st Century: Problems and Frontiers," Moscow, May 13–18, 1995.

Lack, D. *Darwin's Finches*. Cambridge: Cambridge University Press, 1983. First published 1947.

Lane, N. *Life Ascending*. New York: W. W. Norton, 2009.

Lazcano, A., and J. L. Bada. "Stanley L. Miller (1930–2007): Reflections and Remembrances." *Origins of Life and the Evolution of the Biosphere* 38 (2008): 373–81.

Lederberg, J. "Sputnik + 30." *Journal of Genetics* 66, no. 3 (1987): 217–20.

Lederberg, J. "The Transformation of Genetics by DNA." *Genetics* 136, no. 2 (1994): 423–26.

Lederberg, J., and D. B. Cowie. "Moondust." *Science* 127, no. 3313 (1958): 1473–75.

Leeming, D. L., with M. A. Leeming. *A Dictionary of Creation Myths*. Oxford: Oxford University Press, 1994.

"Life and a Glass Earth." *New York Times*, May 17, 1953.

Lindberg, D. "The Fate of Science in Patristic and Medieval Christendom." In *The Cambridge Companion to Science and Religion*, edited by P. Harrison, 21–38. Cambridge: Cambridge University Press, 2010.

Malebranche, N. *De la recherche de la vérité*, vol. 2. Paris: Garnier Frères, 1910.

McCollum, G., and D. Bogard. Interview by N. H. Chaffee, June 18, 2012, transcript, NASA Johnson Space Center Oral History Project.

Medvedev, Z. A. *The Rise and Fall of T. D. Lysenko*. New York: Columbia University Press, 1969.

Middleton, W. E. K. *The Experimenters: A Study of the Accademia del Cimento*. Baltimore: Johns Hopkins University Press, 1971.

Mitford, N. *Voltaire in Love*. New York: Carroll & Graf, 1957.

Moland, L., ed. *Oeuvres complètes de Voltaire*, vol. 14 (Paris: Garnier, 1878).

Morell, V. "Microbiology's Scarred Revolutionary." *Science* 276, no. 5313 (1997): 699–702.

Muller, H. J. "The Development of the Gene Theory." In *Genetics in the 20th Century: Essays on the Progress of Genetics during Its First 50 Years*, edited by L. C. Dunn, 77–99. New York: Macmillan, 1951.

Nagy, B., G. Claus, and D. J. Hennessy. "Organic Particles Embedded in Minerals in the Orgueil and Ivuna Carbonaceous Chondrites." *Nature* 193 (1962): 1129–33.

Nagy, B., and J. J. Lynch, eds. "Life-like Forms in Meteorites and the Problems of Environmental Control on the Morphology of Fossil and Recent Protobionata." *Annals of the New York Academy of Sciences* 108 (1963): 339–616.

Nair, P. "Woese and Fox: Life Rearranged." *Proceedings of the National Academy of Sciences* 109 (2012): 1019–21.

Nichols, A., ed. *Romantic Natural Histories: Selected Texts with Introduction*. Boston: Houghton Mifflin, 2004.

Oren, A., and G. M. Garrity. "Then and Now: A Systematic Review of the Systematics of Prokaryotes in the Last 80 Years." *Antonie van Leeuwenhoek* 106 (2014): 43–56.

Orgel, L. E. "Evolution of the Genetic Apparatus." *Journal of Molecular Biology* 38, no. 3 (1968): 381–93.

"The Origin of Life: Being an Account of Experiments with Certain Superheated Saline Solutions in Hermetically-Sealed Vessels." *Lancet* 1, no. 4675 (1913): 970.

Osborn, H. F. *From the Greeks to Darwin: An Outline of the Development of the Evolutionary Idea*. London: Macmillan, 1905.

Paley, W. *Natural Theology*. New York: American Tract Society, 1881.

Park, D. *The Grand Contraption: The World as Myth, Number, and Chance*. Princeton, NJ: Princeton University Press, 2005.

Parker, G. *The Thirty Years War*. New York: Routledge & Kegan Paul, 1984.

Partington, J. R. *A Short History of Chemistry*. New York: Dover, 1989.

Parton, J. *Life of Voltaire*, vol. 2. Boston: Houghton, Mifflin, 1881.

Pasteur, L. "On Spontaneous Generation: An Address Delivered by Louis Pasteur at the 'Sorbonne Scientific Soiree' of April 7, 1864." *Revue des cours scientifiques*, April 23, 1864, 257–64.

Pearl, D. L. "Political Economy for Workers: A. N. Bakh's *Tsar-Golod*." *Slavic Review* 50, no. 4 (2014): 768–78.

Peretó, J., J. L. Bada, and A. Lazcano. "Charles Darwin and the Origin of Life." *Origins of Life and Evolution of Biospheres* 39, no. 5 (2009): 395–406.

Pinto-Correia, C. *The Ovary of Eve: Egg and Sperm and Preformation*. Chicago: University of Chicago Press, 1997.

Plait, P. "2011 JREF Pigasus Awards." *Discover*, April 1, 2011.

Ponnamperuma, C. "The Origin of Life: From Oparin to the Present." Lecture presented at the International Symposium "Biochemistry of the 21st Century: Problems and Frontiers," Moscow, May 13–18, 1995.

Ponnamperuma, C. *The Origins of Life*. London: Thames and Hudson, 1972.

Poundstone, W. *Carl Sagan: A Life in the Cosmos*. New York: Henry Holt, 1999.

Priestley, J. *The History and Present State of Electricity, with Original Experiments*, 3rd ed. London: C. Bathurst and T. Lowndes, 1775.

Priestley, J. "Observations and Experiments Relating to Equivocal, or Spontaneous, Generation." *Transactions of the American Philosophical Society* 6 (1809): 119–29.

Prum, M. "Charles Darwin's First French Translations." In *The Literary and Cultural Reception of Darwin in Europe*, edited by T. F. Glick and E. Shaffer. New York: Bloomsbury Academic, 2014.

Redgrove, H. S., and I. M. L. Redgrove. *Joannes Baptista van Helmont: Alchemist, Physician and Philosopher*. London: William Rider & Son, 1922.

Redi, F. *Esperienze intorno a diverse cose naturali, e particolarmente a quelle, che ci son portate dall'Indie*. Naples: Giacomo Raillard, 1687.

Redi, F. *Experiments on the Generation of Insects*. Chicago: Open Court, 1909.

Ridley, M. "Crick and Darwin's Shared Publication in *Nature*." *Nature* (2004) 244.

Ridley, M. *Francis Crick: Discoverer of the Genetic Code*. New York: HarperCollins, 2006.

Ridley, M. *Genome*. New York: Harper Perennial, 2006.

Roe, S. "Biology, Atheism, and Politics in Eighteenth-Century France." In *Biology and Ideology from Descartes to Dawkins*, edited by D. R. Alexander and R. L. Numbers, 36–60. Chicago: University of Chicago Press, 2010.

Roe, S. "Voltaire versus Needham: Atheism, Materialism, and the Generation of Life." *Journal of the History of Ideas* 46, no.1 (1985): 65–87.

Roger, J. *The Life Sciences in Eighteenth-Century French Thought*. Stanford, CA: Stanford University Press, 1997.

Rovelli, C. *The First Scientist: Anaximander and His Legacy*. Yardley, PA: Westholme, 2007.

Rupke, N. *Richard Owen: Biology without Darwin*. Chicago: University of Chicago Press, 2009.

Sapp, J. *The New Foundations of Life*. Oxford: Oxford University Press, 2009.

Sapp, J., and G. E. Fox. "The Singular Quest for a Universal Tree of Life." *Microbiology and Molecular Biology Reviews* 77, no. 4 (2013): 541–50.

Sawyer, K. *The Rock from Mars: A Detective Story on Two Planets.* New York: Random House, 2006.

Schopf, W. *Cradle of Life: The Discovery of Earth's Earliest Fossils.* Princeton, NJ: Princeton University Press, 1999.

"Science: Semi-Creation." *Time*, May 25, 1953.

Secord, J. A. "The Curious Case of the *Acarus crossii.*" *Nature* 345 (1990): 471–72.

Secord, J. A. *Victorian Sensation: The Extraordinary Publication, Reception, and Secret Authorship of the Vestiges of the Natural History of Creation.* Chicago: University of Chicago Press, 2000.

Seymour, M. *Mary Shelley.* London: John Murray, 2000.

Shapiro, R. *Origins: A Skeptic's Guide to the Creation of Life on Earth.* New York: Summit, 1986.

Shea, W. R., and M. Artigas. *Galileo in Rome: The Rise and Fall of a Troubled Genius.* Oxford: Oxford University Press, 2003.

Shelley, M. *Frankenstein.* New York: W. W. Norton, 1996.

"Sidney W. Fox; Analyzed First Moon Rocks." *Los Angeles Times*, August 18, 1998.

Soloviev, Y. Y. "240th Anniversary of the Birth of Georges Cuvier (1769–1832)." *Paeleontological Journal* 44, no. 6 (2010): 107–12.

Stott, R. *Darwin's Ghosts.* New York: Spiegel and Grau, 2012.

Strick, J. "Creating a Cosmic Discipline: The Crystallization and Consolidation of Exobiology, 1957–1973." *Journal of the History of Biology* 37 (2004): 131–80.

Strick, J. *Sparks of Life: Darwinism and the Debates over Spontaneous Generation.* Cambridge, MA: Harvard University Press, 2000.

Thomas, S. *Men of Space: Profiles of Scientists Who Probe for Life in Space.* Philadelphia: Chilton, 1963.

Tresch, J. *The Romantic Machine.* Chicago: University of Chicago Press, 2012.

Vallery-Radot, R. *Louis Pasteur—His Life and Labours.* London: Longmans, Green, 1885.

Venter, J. C. *Life at the Speed of Light.* New York: Viking, 2013.

Voltaire. *Oeuvres complètes de Voltaire*, vol. 10. Edited by L. Moland. Paris: Garnier, 1877.

Voltaire. *The Works of Voltaire.* Akron, OH: Werner, 1904.

Voltaire. *The Works of Voltaire*, vol. 33. Paris: A. Firmin Didot, 1829. http://www.gutenberg.org/files/30123/30123-h/30123-h.htm#chap04.

Walcott, C. "Pre-Cambrian Rocks of North America." *Bulletin of the United States Geological Survey*, no. 81 (1891): 594.

Wald, G. "The Origin of Life." *Scientific American* 191, no. 2 (1954): 44–53.

Ward, P. *Life as We Do Not Know It: The NASA Search for (and Synthesis of) Alien Life*. New York: Penguin, 2005.

Watson, J. D. *The Double Helix: A Personal Account of the Discovery of the Structure of DNA*. New York: Simon & Schuster, 1968.

Watson J. D., and F. H. Crick. "Molecular Structure of Nucleic Acids: A Structure for Deoxyribose Nucleic Acid." *Nature* 171 (1953): 737–38.

Wedgwood, C. V. *The Thirty Years War*. New York: New York Review of Books, 2005.

Whitfield, J. "Origin of Life: Nascence Man." *Nature* 459 (2009): 316–19.

Whittaker, E. T. *A History of the Theories of Aether and Electricity from the Age of Descartes to the Close of the Nineteenth Century*. London: Longman, Green, 1910.

Wilson, E. O. *From So Simple a Beginning: The Four Great Books of Charles Darwin*. New York: W. W. Norton, 2006.

Woese, C. "Evolving Biological Organization." In *Microbial Phylogeny and Evolution*, edited by Jan Sapp, 99–118. Oxford: Oxford University Press, 2005.

Woese, C. R. *The Genetic Code: The Molecular Basis for Genetic Expression*. New York: Harper & Row, 1967.

ACKNOWLEDGMENTS

T O TELL A STORY of this scope, we had to draw upon the help and encouragement of countless individuals who provided us with knowledge, insight, and encouragement along the way. Although we have inevitably overlooked some, we'd like to mention a few. Jack Szostak graciously shared his time and opened up his lab at Harvard to prying eyes. George Fox made us laugh with his personal recollections of Carl Woese, too few of which we were able to use. Armen Mulkidjanian of the University Osnabrück helped with research on Alexander Oparin, some of which would have been unavailable to non-Russian speakers. Ron Fox shared personal details about his father, Sidney. Elisa Biondi at the Foundation for Applied Molecular Evolution deserves special praise. By translating for us original source material from Francesco Redi, she went far beyond what we could have reasonably expected.

Merri Wolf and Shawn Hardy of the Geophysical Library of the Carnegie Institution of Washington provided us with access to many hard to find research materials, as only skilled librarians can. We'd also like to thank the Institute for Advanced Studies, without which we would not have been able to access the breadth of Princeton University's library resources.

Gail Ross and Howard Yoon lived up to their reputations as excellent agents. And at Norton, we were lucky enough to cross paths with two wonderful editors—Angela von der Lippe, who believed in this story, and Alane Salierno Mason, whose skill and perserverance made it a reality. Others whose assistance was invaluable include Faye Torresyap, Stephanie Hiebert, and Remy Cawley. And we can't fail to thank our wives, Antje Teegler and

Tracy Wahl, without whom this book would have been impossible. Antje was a diligent and invaluable proofreader, while Tracy provided creative guidance that harnessed years of experience at National Public Radio. Thank you both for putting up with us.

We should point out the obvious fact that neither of us is a historian. The lion's share of the credit for this book is owed to the often underappreciated women and men who have tirelessly preserved the stories of science for posterity. We hope that this book encourages interest in the work of those who have striven to understand not only what we as a society believe, but how and why we came to believe it.

Finally, we'd like to thank those scientists who have devoted and those who continue to devote their lives to humanity's greatest question: How did we get here?

ILLUSTRATION CREDITS

Page 136   Burning barrels of tar to ward off miasmas during the Manchester cholera outbreak of 1832. Wellcome Library, London.

Page 140   Drawing of *Bathybius haekelii* as seen under a microscope. NOAA Central Library Historical Collections.

Page 150   J. B. S. Haldane in 1941. Hans Wild / The LIFE Picture Collection / Getty Images.

Page 162   Charles Doolittle Walcott at the Grand Canyon. Smithsonian Institution Archives. Image # 83-14116.

Page 177   Stanley Miller and his "classical apparatus." © Bettmann / Corbis.

Page 181   Editorial cartoon appearing in the *Washington Post*, December 31, 1956. © A 1956 Herblock Cartoon, © The Herb Block Foundation.

Page 190   Buzz Aldrin drives a core tube sampler into the lunar soil. Photographed by Neil Armstrong. NASA.

Page 205   An iconic 1953 image of Watson and Crick with a DNA double helix model. A. Barrington Brown / Science Source.

Page 215   Possible Martian fossils on ALH84001. NASA / JSC / Stanford University.

Page 225   Stromatolites in Australia's Yalgorup National Park. C. Eeckhout, Creative Commons Attribution 3.0 unported.

Page 232   Woese holds a model of RNA in 1961. Associated Press.

Page 233   (top left) The Tree of Life according to Chambers. CC PD-US. (top right) The Tree of Life according to Woese. NASA. (bottom) The Tree of Life according to Darwin. Wellcome Library, London.

Page 243   A *Tetrahymena* as seen by a scanning electron micrograph. Aaron J. Bell / Science Source.

INDEX

Note: Page numbers in *italics* refer to illustrations.